Visualizing Mathematics with 3D Printing

Visualizing Mathematics with 3D Printing

HENRY SEGERMAN

JOHNS HOPKINS UNIVERSITY PRESS   BALTIMORE

9 8 7 6 5 4 3 2

Johns Hopkins University Press
2715 North Charles Street
Baltimore, Maryland 21218-4363
www.press.jhu.edu

Library of Congress Cataloging-in-Publication Data

Names: Segerman, Henry, 1979–
Title: Visualizing mathematics with 3D printing / Henry Segerman.
Description: Baltimore : Johns Hopkins University Press, 2016. | Includes bibli-
    ographical references and index.
Identifiers: LCCN 2015043848| ISBN 9781421420356 (hardcover : alk. paper) |
    ISBN 9781421420363 (electronic) | ISBN 142142035X (hardcover : alk. paper)
    | ISBN 1421420368 (electronic)
Subjects: LCSH: Geometry—Computer-assisted instruction. | Mathemat-
    ics—Computer-assisted instruction. | Geometry—Study and teaching. |
    Geometrical constructions. | Three-dimensional imaging. | Three-dimen-
    sional printing.
Classification: LCC QA462.2.C65 S44 2016 | DDC 516.028/6—dc23 LC record
    available at http://lccn.loc.gov/2015043848

A catalog record for this book is available from the British Library.

*Special discounts are available for bulk purchases of this book. For more informa-
tion, please contact Special Sales at 410-516-6936 or specialsales@press.jhu.edu.*

Johns Hopkins University Press uses environmentally friendly book materials, including recycled text paper that is composed of at least 30 percent
post-consumer waste, whenever possible.

Contents

Preface

Welcome to my book, dear reader. Before anything else, let me first encourage you to visit the companion website to this book, 3dprintmath.com.

This is a popular mathematics book, intended for everyone, no matter his or her mathematical level. This book is a little different from other popular math or science books. In this book, whenever it makes sense, the diagrams are photographs of real-life 3D printed models. Almost all of these models are available virtually on the website—they can be rotated around on the screen so that you can view them from any angle. They are also available to download and 3D print on your own 3D printer or purchase online at the website.

With these models, you, the reader, can experience three-dimensional concepts directly, as three-dimensional objects. They let me describe some very beautiful mathematics, including some topics that, although accessible, are difficult to explain well using only two-dimensional images. I've tried hard to make things understandable with only the book, but ideally you should be reading while holding the 3D printed diagrams in your hands or using the virtual models on the website.

Because this book is built around 3D printed diagrams, the topics we will look at tend toward the geometric. The first chapter is about the different ways that three-dimensional objects can be symmetric. Chapter 2 is about some of the simplest shapes: the two-dimensional polygons and the polyhedra, their three-dimensional relatives. Chapter 3 builds off chapter 2, reaching up to the four-dimensional relatives of polygons and polyhedra and investigating how we can see four-dimensional objects by casting shadows of them down to three dimensions. Chapter 4 is about tilings and curvature—whether a surface is shaped like a hill, a flat plane, or a saddle. Chapter 5 is about knots and thinking topologically—looking at geometric objects but not caring about the precise shapes, as if everything were made of very stretchy rubber. Chapter 6 continues the topological theme by looking at surfaces and then later on thinking about geometry again by putting tilings on surfaces. Chapter 7 is a menagerie, of mathematical prints I couldn't resist including in the book.

Appendix A lists credits and technical details for the figures and 3D prints. Some of these include parametric equations that the adventurous reader may want to use to create her own visualizations. If you're interested in how I go about making models, see appendix B.

There isn't much in the way of tricky notation or calculations in this book. It's more about getting a visual sense of what's going on. Having said that, some things might be a bit difficult to follow. If you get stuck on something, feel absolutely free to skip over it and come back later.

Why 3D printing? 3D printing is a technology that gives unprecedented freedom in the creation of three-dimensional physical objects. A 3D printer builds an object layer by layer in an automated additive process, based on a design given to it by a computer. 3D printers are particularly suited to producing mathematically inspired objects, in part because designs can be generated by programs written to precisely represent the mathematics. Because production is automated, the physical models you obtain closely approximate the mathematical ideals. Small production runs of 3D objects and production on demand aren't as possible with other manufacturing technologies. There's no way I could have made all of the diagrams in this book without 3D printing.

One last comment before we get started: 3D printers are so good at producing mathematical models that I sometimes run into an interesting problem. A photograph of a physical 3D printed object is so close to the mathematical ideal that viewers sometimes assume that the photograph is actually a computer render. All of the pictures in this book that look like photographs of real objects are indeed photographs of real objects, sometimes with some color added to the image to highlight a feature. I have deliberately left occasional imperfections in the photographs to prove their reality. Or at least, this seems like an excellent excuse/reason for any flaws you may find.

Acknowledgments

This book would not have happened without many, many other people. First of all, my parents, Eph and Jil Segerman, were instrumental in my existence. Apparently, they also had a large part to play in getting me into 3D stuff in the first place, because both my brother, Will, and I have (in very different ways) built our careers around 3D. Will's current main source of income is as a virtual milliner. Make of that what you will.

Huge thanks to my various collaborators. My brother, Will, worked with me on the monkey sculptures, and Vi Hart got us thinking about four-dimensional symmetries. Keenan Crane flowed a coffee mug, Geoffrey Irving figured out where to put hinged triangles, Craig S. Kaplan tiled a bunny, Marco Mahler worked with me on mobiles, and Roice Nelson tiled two- and three-dimensional hyperbolic space. Particular thanks to Saul Schleimer, my collaborator in both topology research and mathematical illustration, who is very easy to distract from the former to the latter. Saul and I worked on too many projects to list here, apart from the one with yet another collaborator: the parametrization of the figure-eight knot, with François Guéritaud.

Thanks to the other mathematicians, designers, and artists whose work I featured: Vladimir Bulatov, Bathsheba Grossman, George Hart, Oliver Labs, Carlo Séquin, Laura Taalman, Oskar van Deventer; Jessica Rosenkrantz and Jesse Louis-Rosenberg of Nervous System; the team that worked on the ropelength knots: Jason Cantarella, Eric Rawdon, Michael Piatek, and Ted Ashton; and the team that worked on the flat torus: Vincent Borrelli, Saïd Jabrane, Francis Lazarus, and Boris Thibert.

Thanks to Bus Jaco (the head of) and the rest of the Department of Mathematics at Oklahoma State University, for their support while I was writing the book. Bus helped me track down the OSU physics and chemistry instrument shop that built the photo rig for me and also found Joyce Lucca and Sam Welch, who loaned me the turntable. The purchases of many of the models were supported by a Dean's Incentive Grant from the College of Arts and Sciences at OSU.

Thanks to Robert McNeel & Associates for making Rhinoceros, the main program I used to design the models, and to the 3D printing service Shapeways for printing them.

Jarey Shay designed the companion website to the book, and NeilFred Picciotto acquired the domain names.

Stephan Tillmann and an anonymous reviewer both made early suggestions that changed the core focus of the book. I had some useful conversations about negatively curved spaces with Chaim Goodman-Strauss.

Thanks to Vincent J. Burke, Andre M. Barnett, and everyone else at Johns Hopkins University Press, who turned my manuscript into a book.

Moira Bucciarelli, Evelyn Lamb, Craig Kaplan, Rick Rubinstein, Saul Schleimer, Jil Segerman, Carlo Séquin, Rosa Zwier, and the anonymous reviewers read through versions of the book and found lots of ways to make it better and clearer. All errors are, of course, my own.

Visualizing Mathematics with 3D Printing

1 Symmetry

Symmetrical objects and patterns surround us, in art, architecture, and design. We are mostly symmetrical, at least on the outside. What is symmetry? How can we recognize different kinds of symmetry?

A *symmetry* of an object is a *motion of the object that leaves it looking the same*. There are eight motions that leave the design in fig. 1.1 looking the same, and so there are eight symmetries: We can rotate by one-eighth of a turn, two-eighths, three-, four-, five-, six-, or seven-eighths of a turn. I also want to count the "do nothing" motion, in which we don't do anything at all.

Fig. 1.2 shows a different kind of symmetry. Here there are rotations that leave the design looking the same, but there are also reflections. Let's also think of reflections as motions, so that this design has eight symmetries: rotating by one-quarter of a turn, two-quarters of a turn, and three-quarters of a turn, the do-nothing motion, and the reflections in the four red lines.

These are both examples of two-dimensional symmetrical designs in the plane—they are flat, printed on a page of this book. Since this is a book about 3D printed things, we'll mostly look at symmetries of three-dimensional objects. Rather than jumping

Fig. 1.1. A design with rotational symmetries.

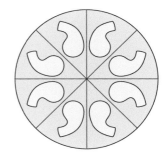

Fig. 1.2. A design with both rotational and reflectional symmetries.

straight into talking about what the different symmetries of a three-dimensional object can be, let's take a scenic route, via a different interesting question.

What Are the Different Ways to Take a Photograph of a Three-Dimensional Object?

Let's start with a very simple object—a sphere. Fig. 1.3 is a photograph of a spherical bubble. How many different ways could I have taken this photograph? Let's set some rules for my photography. Suppose the following hold true:

1. I take all of my photographs from the same distance, pointing directly at the object I'm photographing.
2. I don't care about changes in lighting and shadows, only the shape I see.

Fig. 1.3. A spherical bubble.

Fig. 1.4. A bottle.

The only thing that matters is the direction from which we take the photograph. Well, a sphere looks the same from every direction. So, as far as I'm concerned, there is *only one* photograph of a sphere. It always looks the same, no matter which angle you look at it.

Next, let's consider a more complicated object—a bottle. See fig. 1.4. Now, the direction that I take a photograph from matters. Well, sometimes it matters. If I walk around the bottle taking photographs, keeping my camera at the same height and always pointing at the bottle, then I'll always get the same picture. But if I move the camera up or down, the photograph I will change. See fig. 1.5.

What's going on here is that there is a sphere of possible directions from which to take a photograph. Rotating around the bottle doesn't change anything—only moving up or down creates different photographs. We only need to take photographs along a semicircle to get all possible photographs. See fig. 1.6.

If we move off this semicircle to the side, then we just see the same thing again. Is there anything else I could do? I could also rotate the camera by "rolling"

Fig. 1.5. *Left,* Different photographs of a bottle.

Fig. 1.6. *Above,* A semicircle of camera positions.

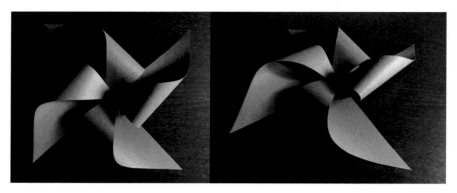

Fig. 1.7. A paper windmill.

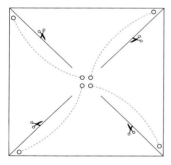

Fig. 1.8. To make a paper windmill: cut a square of paper along the lines from the corners toward the center and then glue the corresponding spots together.

it to the side, without changing the direction it is pointing in. Then, the photographs I take would be rotated versions of one another. This doesn't really show us anything new. In fact, let's add this as a third rule:

3. Photographs that are rotations of one another are really the same.

This means that the sphere of possible directions from which to point the camera at the bottle is all that matters.

Next, let's think about a paper windmill. See fig. 1.7. In case you don't have one of these handy, you can make one from a square of paper (see fig. 1.8). Suppose that the windmill is lying down flat, with the front pointing upward. If we move our camera up and down along the same semicircle of viewpoints we used for the bottle, then we will again see a different photograph from each viewpoint. This time, if we move off to the side, we get new views, which are different from any we have seen on the semicircle. Once we have moved around by a quarter of a turn, we start seeing the same photographs again. Instead of the semicircle for the bottle, we get

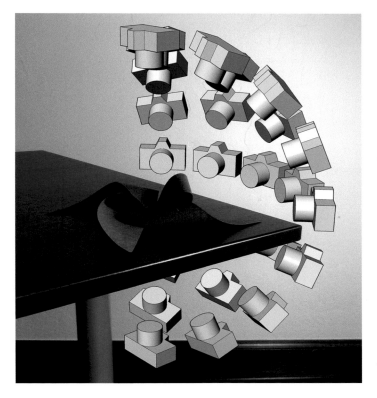

Fig. 1.9. The panel of possible views of a paper windmill.

a quarter of the sphere: one of the four panels of a beach ball. Fig. 1.9 shows one of these panels of views. Within any one panel, every viewpoint gives a different photograph, but if we move to a viewpoint in another panel, then we see the same photographs again.

Now something a little more complicated, although, at first, it might seem like a simpler thing—a cube. See fig. 1.10. What are the different ways to photograph a cube? This is tricky.

Get a cube to look at. You may think you already know what a cube looks like, but it will help for the next bit to get an actual cube to look at from different directions. Even if you don't have a 3D printed cube at hand, you probably have something nearby. A sugar cube, a six-sided die, a Rubik's cube, or a Minecraft block? I'll wait.

Now, you have your cube. It's like the paper windmill in that you can rotate it by a quarter of a turn to the side and it looks the same. But it also looks the same after rotating it upward by a quarter of a turn. If you hold it by opposite corners as on the right of

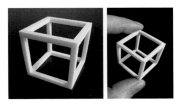

Fig. 1.10. A cube, 3D printed in nylon plastic.

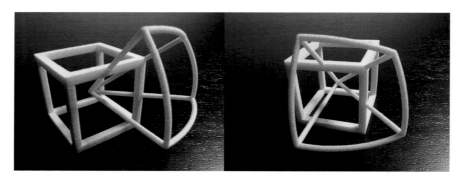

Fig. 1.11. All of the possible views of a cube can be seen through this kite-shaped panel.

fig. 1.10 and spin it between your finger and thumb, it looks the same after rotating by a third of a turn. There are more ways to move it and have it look the same than for the paper windmill, so less of the sphere of possible views of the cube actually consists of different views. The panel of different views for the cube is smaller: rather than the beach ball panel we got for the windmill, we get a kite-shaped panel for the cube. See fig. 1.11.

Fig. 1.12 shows some of the actual photos you get when pointing at the cube from the directions in this kite. Fig. 1.13 shows the kite-shaped panel again, but with little camera models to represent the position from which I photographed the cube. (Last chance if you still haven't found a cube to look at. You can also rotate a 3D model around on the website; see 3dprint math.com.)

I took this grid of photographs using a rig that allows me to (relatively) precisely control the angle that the camera sees the cube from (see fig. 1.14). With this setup, the camera is fixed while the cube can be rotated in various ways, but it is probably easier to think about this as we have been doing—moving the camera around the fixed cube.

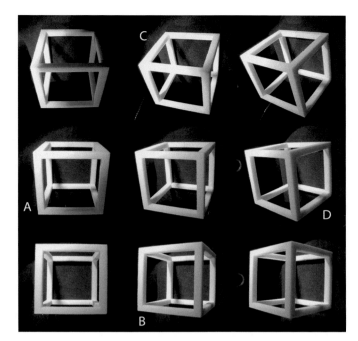

Fig. 1.12. *Top,* Nine ways to take a photograph of a cube.

Fig. 1.13. *Middle,* Camera positions for the nine views of a cube in fig. 1.12.

Fig. 1.14. *Bottom,* How to take photographs from any direction.

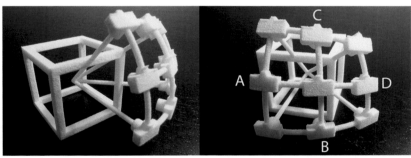

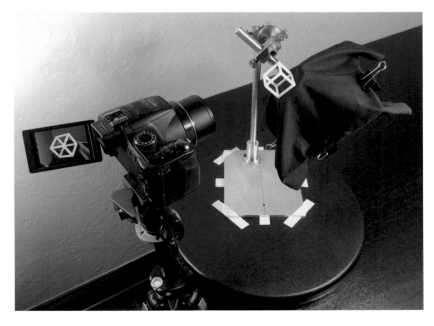

How can we be sure that this kite-shaped panel gives us all of the different photographs of a cube? Could we have missed some? Look at view A in figs. 1.12 and 1.13. View A looks the same as view B. It is rotated, but we decided in rule 3 to think of rotated photographs as being the same. Now, what is the view farther to the left side of view A in fig. 1.12? That is, what would you see if you rotated your head around the cube a little to the left from view A? Well, A is the same as B. Turn the book by a quarter of a turn counterclockwise, and view B matches up exactly with view A from before you turned the book. The view a little to the left from this rotated view B is already printed on the page. It's the view in the center of the grid of photographs.

Said another way, if we rotate our point of view from A off to the left, outside of the grid of photographs shown in fig. 1.12, what we see is the same as rotating our view from B upward, into the grid of photographs, and we already have those.

To the left of A is another copy of our grid of photographs, which we can represent by adding a new kite-shaped panel around our cube, as shown in fig. 1.15. Anything we can see by looking into the second panel we can also see by looking into the first.

This is what symmetries are all about. An object looks the same when you move it around to a different position from where you started or, equivalently, when you move your camera around it to look at it from different directions.

The C and D views in figs. 1.12 and 1.13 are also the same, and so again, there is another copy of our grid of photographs and another panel of views above C. Carrying on like this, we can cover the whole sphere of possible photographs, with 24 copies of the panel (four for each of the six faces of the cube), as shown in fig. 1.16. So one panel really gives us all of the possible photographs of a cube: every direction we could look at the cube is covered by some copy of the panel.

You might have noticed that there is a line of mir-

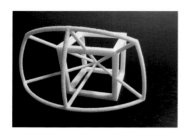

Fig. 1.15. Two kite-shaped panels.

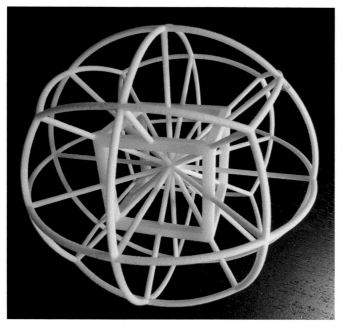

Fig. 1.16. Twenty-four kite-shaped panels.

ror symmetry in fig. 1.12. Photographs on either side of the diagonal line from bottom left to top right can be reflected onto one another. Maybe we are double counting the different ways to take a photograph?

So far, we have been treating two photographs as the same if one is a rotation of the other. But we could add a fourth rule:

4. Photographs that are reflections of one another are really the same.

With this new rule, the photographs in the top left triangle are the same as photographs in the bottom right triangle, and we can cut our kite-shaped panel of possible photographs in half, as in fig. 1.17. This means that the 24 kite-shaped panels that cover the sphere are cut into 48 copies of this smaller triangular panel, as in fig. 1.18.

The model in fig. 1.18 is complicated, and it can be difficult to see what is going on. Fig. 1.19 shows an alternative model with the same information. This model is called *Comma symmetry sphere *432*. I'll come back in a little while to explain this somewhat cryptic name. The comma design is repeated 48 times over the surface of the sphere, once for each triangular

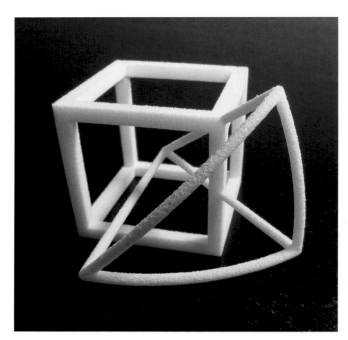

Fig. 1.17. A half-kite-shaped panel.

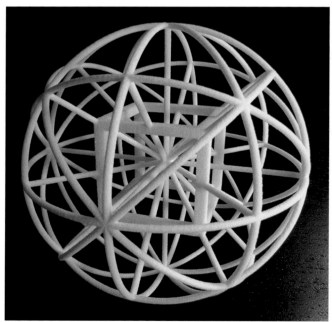

Fig. 1.18. Forty-eight half-kite-shaped panels.

panel in fig. 1.18. The symmetries of the cube mean that you get a different picture for every view within a panel, but they repeat if you move to a different panel. The commas repeat in exactly the same way, and their shape makes it easy to see how they are arranged on the surface of the sphere.

Pick your favorite comma, and call it the *home com-*

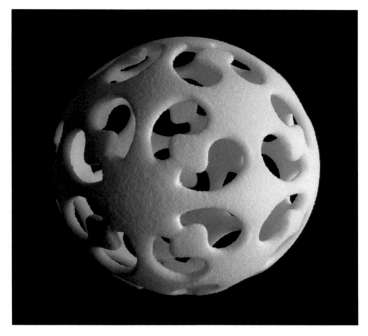

Fig. 1.19. *Comma symmetry sphere *432.*

ma. For every other comma, there is a motion of the model (remember that we are also thinking of reflections as motions) that takes it to the home comma. Including the do-nothing motion, there are a total of 48 possible motions, and there are a total of 48 symmetries of this comma sphere.

This is a lot of symmetries, although that shouldn't be too surprising because the cube is a very symmetrical object.

These rotations and reflections are the key to understanding the symmetries of three-dimensional objects. Any symmetrical object (e.g., the sculptures in figs. 1.30 to 1.37) has an underlying symmetry that can be represented by a comma sphere (see fig. 1.19) and by a notation for its symmetry type such as *432). Before getting into the notation, let's look at another example. Fig. 1.20 shows two photographs of *Soliton*, a sculpture by mathematical artist Bathsheba Grossman.

This is a difficult object to comprehend from a few photographs. Sinuous curves twist around one another in a complicated, but obviously symmetrical, way. Rotation by half a turn is a symmetry for each of these views. But it isn't so easy to see how these two views are related to each other or even that they are pho-

Fig. 1.20. *Soliton* by Bathsheba Grossman.

Fig. 1.21. Many ways to take a photograph of *Soliton*. Try getting a stereo vision effect by looking at two neighboring photographs with different eyes.

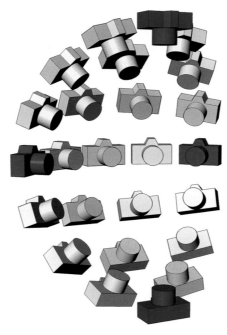

Fig. 1.22. One "beach ball panel" for *Soliton*.

tographs of the same object. With a few more view-points of the same sculpture, however, we can see how they are connected. See fig. 1.21. The first view shown in fig. 1.20 is at the far right, and the second is at both the top and the bottom.

Again, let's think of the sculpture sitting at the center of a sphere of possible directions from which to take a photograph. This time our panel is a quarter of the entire sphere, like the panel of a four-panel beach ball. This panel has the same shape as the panel for the paper windmill but beware—it doesn't have the same symmetries. One of the symmetries of the wind-mill rotates it by a quarter turn, while *Soliton* only has rotations by half a turn.

Fig. 1.22 shows camera positions evenly spaced out over one of the panels. Photographs from these posi-tions make up fig. 1.21.

As with the cube, some of the photographs around the edges are repeats: they show the same view as each other (remember rule 3 says that rotated photographs are the same as one another). The pair of photographs above and below the rightmost photograph in fig. 1.21 are the same as each other, as are the pair two above and two below, and so on. In fact, the whole boundary edge from the rightmost point to the top is the same as the edge from the rightmost point to the bottom. The same is true of the two edges above and below the leftmost edge. This tells us how to cover the rest of the sphere of possible photographs: we can do this with a total of four copies of the panel, tiling the sphere so that the photographs we see along the edges match up.

Looking ahead, *Soliton*'s symmetry type is 222, which is a member of the infinite family 22N.

What Are the Different Ways in Which an Object Can Be Symmetrical?

In this book, we will be mostly thinking about the symmetries of three-dimensional objects. In other words, we are considering *spherical symmetry*, the kinds of symmetrical pattern that could be drawn on the surface of a sphere. This matches with the symmetries of objects that you can pick up and turn around in your hands. At the start of the chapter, we looked at a couple of examples of objects with *rosette symmetry* (see figs. 1.1 and 1.2). The design is in the plane, and all the symmetries fix one central point. There are other kinds of symmetry in the plane, for example, the symmetries of a tiled floor (*planar symmetry*), which also involve *translations*, motions that slide the tiling along the floor without rotating or reflecting it. We won't look closely at the kinds of symmetries other than spherical symmetries. If you're interested in further reading about symmetry, I recommend the wonderful book *The Symmetries of Things* by John H. Conway, Heidi Burgiel, and Chaim Goodman-Strauss, which also details the notation for symmetries I'm about to describe in this chapter and explains why it works.

We have now seen five different kinds of symmetrical object: (1) the sphere, (2) the bottle, (3) the windmill, (4) the cube, and (5) *Soliton*. These all seem to have different kinds of symmetry, but how can we tell, and what are the other possibilities?

The sphere and the bottle both have an infinite number of symmetries. For the sphere, *any* rotation or

reflection is a symmetry. For the bottle, the symmetries are rotations around the vertical axis and reflections in vertical planes that contain that axis. Let's set
these aside, and only look at objects that have a finite
number of symmetries. It turns out that these can be
classified in terms of certain *features* of a symmetrical
object.

When looking at a symmetrical object, the first
kind of feature to look for is any viewpoints from
which the a photograph of the object has rotational
symmetry. For *Soliton*, these are the viewpoints at
the far left, right, and the viewpoint at the top (which
is the same as the viewpoint at the bottom) of fig.
1.21. Each viewpoint gets a number associated with
it. Here, all three viewpoints get a 2, because they all
have twofold rotational symmetry. These views look
the same if you turn the book upside down.

Over the whole sphere of possible viewpoints to
look at *Soliton*, there are a total of six different viewpoints with rotational symmetry. Four of these are
shown in blue in fig. 1.22, and there are two more
around the equator of the sphere. However, we only
count three different *kinds* of viewpoint because there
is a symmetry of the whole object that takes each
viewpoint to its pair on the opposite side of the object.
There are no other symmetrical features, and so the
symmetry type of *Soliton*, is 222, one number for each
kind of symmetric viewpoint. This is three copies of
the number 2, so it is pronounced two-two-two. It
isn't the number two hundred and twenty two.

Fig. 1.23 shows another example, *Comma symmetry
sphere 88*.

This is a little difficult to see in the photograph
on the left (it's easier to see with a 3D model) so I've
added a schematic diagram of the shape on the right.
From now on, I'll show the position of any rotation
axes with small circles.

There are two different viewpoints of *Comma
symmetry sphere 88*, which have eightfold rotational
symmetry, one at the top and one at the bottom, and
so its symmetry type is 88, hence the name. Unlike for

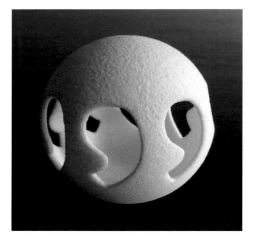

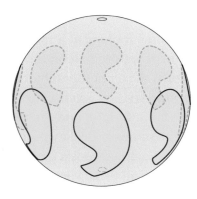

Fig. 1.23. *Comma symmetry sphere 88.*

Soliton, the top view and the bottom view are different, so these each get counted.

The second kind of feature is mirror planes. See, for example, *Comma symmetry sphere *44*, as shown in fig. 1.24. An asterisk (*) in a symmetry type, as in *44, denotes the presence of mirror symmetry in an object.

Here four mirror planes (shown by raised lines) meet along the north-south axis of the sphere, so we say that the two viewpoints on this axis have fourfold *kaleidoscopic symmetry*. We saw the same thing in the design in fig. 1.2. Two different viewpoints of *Comma symmetry sphere *44* have fourfold kaleidoscopic symmetry (one at the top and one at the bottom), but there are no other features, so the symmetry type of this object is *44.

Kaleidoscopic viewpoints also have rotational symmetry, but if a viewpoint has any reflections, then we always call it kaleidoscopic rather than rotational. An object can have both purely rotational and reflectional features. For example, *Comma symmetry sphere 8** (fig. 1.25) has one unique viewpoint with eightfold rotational symmetry and a single mirror plane. We only count one rotational symmetry viewpoint, because the south pole and north pole views are reflections of each

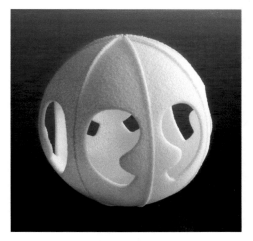

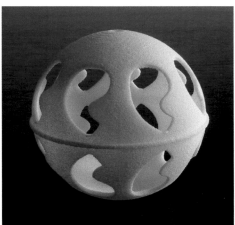

Fig. 1.24. *Top, Comma symmetry sphere *44.*

Fig. 1.25. *Bottom, Comma symmetry sphere 8*.*

other. Similarly, *Comma symmetry sphere 2*4* (fig. 1.26) has one view with twofold rotational symmetry and one with fourfold kaleidoscopic symmetry. Numbers after an asterisk (*) denote points with kaleidoscopic symmetry, while those before (if there are any) denote purely rotational symmetry.

Finally, there is one further kind of symmetrical feature that can arise. Here, a path can be drawn from a comma to a mirror-image comma but without crossing a mirror plane. This "sliding reflection" is denoted by "×". This happens on *Comma symmetry sphere 8×* (see fig. 1.27).

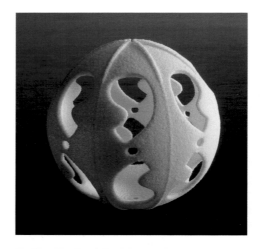

Fig. 1.26. *Comma symmetry sphere 2*4.*

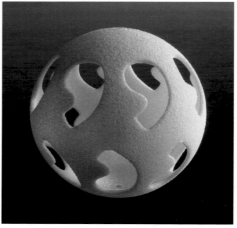

Fig. 1.27. *Comma symmetry sphere 8×.*

Now we can list all of the possible ways to be symmetrical on a sphere. There are seven infinite families and seven more oddities that don't fit into the families. This isn't unusual in mathematics: when organizing things into categories, it very often happens that there are some nice, orderly, infinite families and then a finite (usually small) number of oddities. Fig. 1.28 shows examples of each of the seven infinite families, and fig. 1.29 shows the seven oddities.

We have already seen many of these examples, and we can imagine altering an example to get others in its infinite family. For example, we could change 88 into

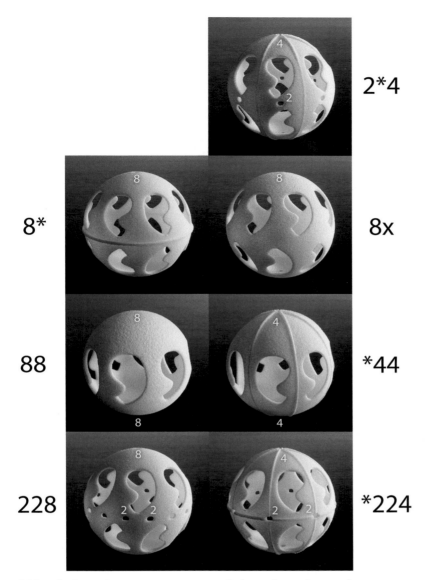

2*4

8*   8x

88   *44

228   *224

99 by fitting nine commas around the sphere instead of eight.

Fig. 1.28. Examples of the seven infinite families.

The infinite families are 2*N, N*, Nx, NN, *NN, 22N, and *22N. For these, N can be any positive integer. We even allow N to be one, and omit any digits "1" that we see. So, really, 1* = * = *11. This makes sense since a point with onefold rotational symmetry doesn't really have any symmetry at all, and a point with onefold kaleidoscopic symmetry has only one mirror line passing through it. The seven oddity symmetry types are 3*2, 332, *332, 432, *432, 532, and *532.

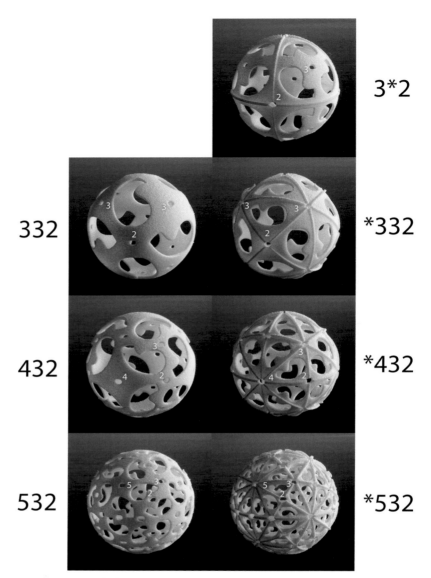

Fig. 1.29. The seven oddities.

Let's look at some beautiful sculptures by mathematical artists Bathsheba Grossman (figs. 1.30, 1.31, and 1.32), George Hart (figs. 1.33 and 1.34), and Vladimir Bulatov (figs. 1.35 and 1.36). Try to figure out the symmetry type of each one. Don't worry if you have trouble with this. It's very difficult. Answers are in appendix A.

Fig. 1.30. Three views of *Double Zarf*, by Bathsheba Grossman.

Fig. 1.31. Two views of *Tentacon*, by Bathsheba Grossman.

Fig. 1.32. Three views of *Metatrino*, by Bathsheba Grossman.

Fig. 1.33. *Six Nested Truncated Cuboctahedra Centerpiece*, by George W. Hart.

Fig. 1.34. *Solar Centerpiece*, by George W. Hart.

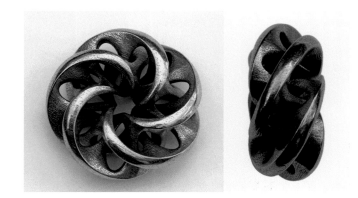

Fig. 1.35. Two views of *Moebius II*, by Vladimir Bulatov.

Fig. 1.36. Two views of *Rhombic Triacontahedron IV*, by Vladimir Bulatov.

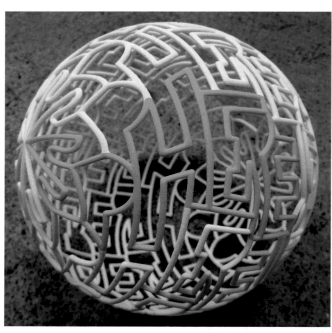

Fig. 1.37. *"Sphere" Sphere*.

One last example: fig. 1.37 shows a symmetric self-referential sculpture. Again, what is the symmetry type (ignoring the coloring)?

2 Polyhedra

You may have noticed that the seven "oddity" symmetry types shown in fig. 1.29 have a certain polyhedral feel to them. This isn't a coincidence because like the cube we started with, the regular polyhedra all have these odd symmetry types. But I'm getting ahead of myself. Just what is a polyhedron?

Polyhedra are the three-dimensional versions of *polytopes*. Let's start with the simplest possible polytope:

A *zero-dimensional polytope* is just a point, or a vertex.

A *one-dimensional polytope* is a line segment, or an edge. That is, it is part of a line, bounded by a pair of zero-polytopes, one at each end.

A *two-dimensional polytope* is a polygon, which is part of a plane, bounded by some number of one-polytopes. For example, a square is a polygon, bounded by four line segments.

A *three-dimensional polytope* is a polyhedron, which is part of three-dimensional space, bounded by some number of two-polytopes. For example, a cube is a polyhedron, bounded by six squares. See fig. 2.1.

In terms of symmetry, we are really interested in the *regular* polytopes. A regular polygon has all of its

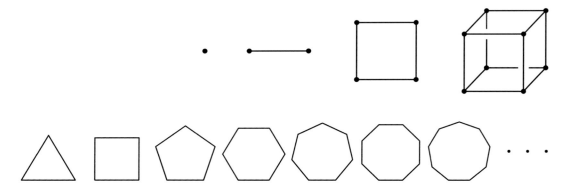

sides the same length and all of the angles at the corners the same. So the regular polygons are the equilateral triangle, the square, the regular pentagon, and so on. See fig. 2.2. A polygon that isn't regular either has sides of different lengths (e.g., a rectangle), corners with different angles (e.g., a rhombus), or both.

It's a little trickier to write down what it means for a polyhedron to be regular. For example, it isn't quite enough to say that we want its faces to be regular polygons, with the same number around each vertex: The "dented" icosahedron on the left of fig. 2.3 has five regular triangles around each vertex, and it certainly isn't very regular.

However, there is a nice way to say what "regular" means that works for polytopes in every dimension. For this, we need to know what a *flag* of a polytope is.

For a three-dimensional polytope (i.e., a polyhedron), a flag consists of a vertex, with an edge that touches the vertex, a face that touches both the vertex and the edge, and the polyhedron itself (that touches the vertex, the edge, and the face). See fig. 2.4.

For a two-dimensional polytope (i.e., a polygon), a flag has only a vertex, an edge, and the polygon. In general, a flag has "sides" of all dimensions from

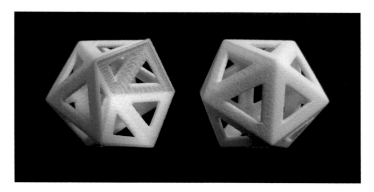

Fig. 2.3. A dented icosahedron, next to an icosahedron.

Fig. 2.4. A flag of a cube. The vertex is marked in red, the edge in green, the face in blue, and the polyhedron in gray.

zero up to the dimension of the polytope itself. Now we can say what "regular" really means: a polytope is *regular* if there is a symmetry (i.e., a motion) of the whole polytope that takes any flag to any other flag.

You might want to stop for a minute to convince yourself that our definition of a regular polygon matches the previous definition of all edges being the same length and all angles at the corners being the same. For example, fig. 2.5 shows the eight different flags of a square. Because the square is regular, for each pair of flags there is a symmetry (in fact, there is only one symmetry) that takes one flag to the other. A symmetry is a way to move the square around so that it looks the same after you're done moving it. To show that some pair of edges of the square are the same length, find the symmetry that moves one edge to the other. Because the square looks the same before and after, despite having moved one edge to where the other was before, the two edges must have the same length. A similar argument shows that all of the angles also have to be the same.

For regular polytopes, flags act like commas on the comma symmetry spheres in chapter 1. We can use them to count how many symmetries there are. In our

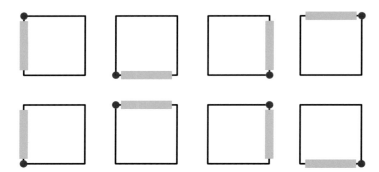

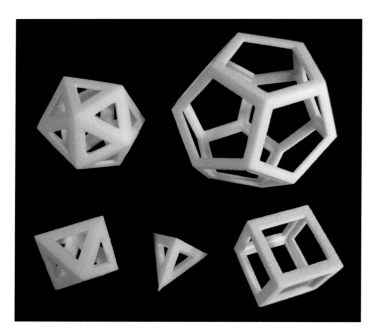

Fig. 2.5. *Top,* The eight different flags of a square. A flag of a square resembles a real-life flag flying on a flagpole.

Fig. 2.6. *Bottom,* The regular polyhedra. *Clockwise from top left*: the icosahedron, dodecahedron, cube, tetrahedron, and octahedron.

example, the square has eight symmetries, matching the eight flags.

There are infinitely many regular polygons because we can just keep adding sides as in fig. 2.2. But in three dimensions, there are only five regular polyhedra: (1) the tetrahedron, (2) the cube (also known as

the hexahedron), (3) the octahedron, (4) the dodecahedron, and (5) the icosahedron. See fig. 2.6.

The tetrahedron has symmetry type *332, the cube
and octahedron both have symmetry type *432, and
the dodecahedron and icosahedron both have symmetry type *532.

Why Are These Five the Only Regular Polyhedra?

Because you can take any flag of a regular polyhedron
to any other by symmetries, a regular polyhedron
must have the following three properties:

1. Its faces must be regular polygons.
2. The faces must all be copies of the same regular
   polygon.
3. There must be the same number of faces around
   each vertex.

The first property is true because we can look at
the flags with one particular face of the polyhedron as
their face. If we throw away the polyhedron from each
of those flags, then we get the set of flags of the face
itself. And we can take any of those to any other, which
is exactly what it means for the face to be a regular
polygon. For the second property, there are symmetries that take any face to any other face because the
faces are parts of the flags. So all of the faces must be
the same. And for the third property, there are symmetries that take any vertex to any other vertex because
the vertices are parts of the flags. All of the vertices are
the same, which means they all have the same number
of faces around them.

The only choices are in which polygon we use for
the faces, and how many of that polygon go around
each vertex. See fig. 2.7.

First, suppose that all of the faces are regular triangles. We must have at least three faces around a vertex,
because two faces glued together around a vertex
would be glued flat against each other. There wouldn't
be any three-dimensional space contained inside the
polyhedron. Three triangles meeting at a vertex give

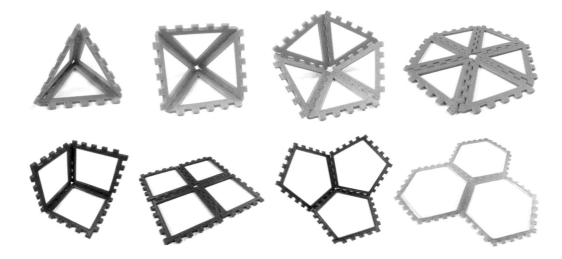

Fig. 2.7. Polygons fitting around a single vertex.

us the tetrahedron, four give us the octahedron, and five give us the icosahedron.

The *angle defect* at a vertex of a polyhedron is how much less than the full 360 degrees you get by adding up the angles of the faces at that vertex. In other words, it measures how far away from being flat the polyhedron is at this vertex. For every vertex of the tetrahedron, the angle defect is $360 - 3 \times 60 = 180$, because there are three triangles at the vertex, each of which contributes 60 degrees. For the octahedron, the angle defect is $360 - 4 \times 60 = 120$, and for the icosahedron, we get $360 - 5 \times 60 = 60$. If we fit six equilateral triangles around a vertex, then the angles add up to exactly 360 degrees, and the angle defect is zero—it's flat. With fewer triangles, the defect is positive, and the faces curve around to contain a region of space. With six triangles, the angle defect is not positive and the faces don't curve around, so we don't get a polyhedron.

Next, three squares around a vertex makes the cube, with defect $360 - 3 \times 90 = 90$, but four around a vertex has defect zero. Three pentagons around a vertex makes the dodecahedron, with defect $360 - 3 \times 108 = 36$, but more than three would have negative defect. And that's it. Three hexagons at a vertex have defect zero already, and three regular polygons with more than six sides would have negative defect, so there are no more regular polyhedra. Our list is complete.

Schläfli Symbols

The first part of this argument shows that, once we know which polygon we are using and we know how many of them go around a vertex, then we know which polyhedron they make. For example, we could write {4,3} for the cube, because it is made out of squares (four sides), with three around each vertex, and {5,3} for the dodecahedron, because it is made out of pentagons (five sides), with three around each vertex. These are the *Schläfli symbols* (named after the Swiss mathematician Ludwig Schläfli) for the cube and dodecahedron. There are also Schläfli symbols for the regular polygons ({3} for the triangle, {4} for the square, and so on), and for higher-dimensional polytopes (coming up in chapter 3).

Positive angle defect corresponds to the faces curving around to enclose a region of space, so we get a polyhedron, and zero angle defect corresponds to the faces forming a flat plane. Can we make sense of negative angle defect? Yes, we can (see in chapter 4). For now, let's notice an amazing coincidence (which, of course, isn't a coincidence at all), written in the form of table 2.1.

Let's finish off this chapter by looking a little at what happens when we relax the rules for making regular polyhedra. There are many ways to do this, but one of the most interesting is to say that the faces have to be regular polygons, but they don't all have to be the same, and that there are symmetries of the polyhedron to take any vertex to any other. Now what is possible?

We get two infinite families, and 13 oddities. The two infinite families are the *prisms* and the *antiprisms* (see fig. 2.8). A prism has two polygons, one at each end, connected by a ring of squares. An antiprism has

Table 2.1 Angle defects for the regular polyhedra

Polyhedron	Schläfli symbol	Angle defect	No. vertices	Defect × No. vertices
Tetrahedron	{3,3}	180	4	720
Cube	{4,3}	90	8	720
Octahedron	{3,4}	120	6	720
Dodecahedron	{5,3}	36	20	720
Icosahedron	{3,5}	60	12	720

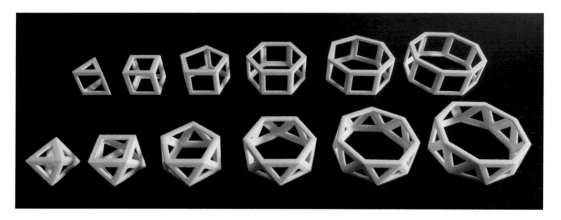

Fig. 2.8. Prisms and anti-prisms.

the two polygons again, but they are twisted relative to each other and connected by two rings of triangles. This doesn't always give us new things. The square prism is the cube again, and the triangular antiprism is the octahedron again.

The 13 oddities are called the *archimedean polyhedra*. See fig. 2.9. They are

1. truncated tetrahedron
2. cuboctahedron
3. truncated cube
4. truncated octahedron
5. rhombicuboctahedron

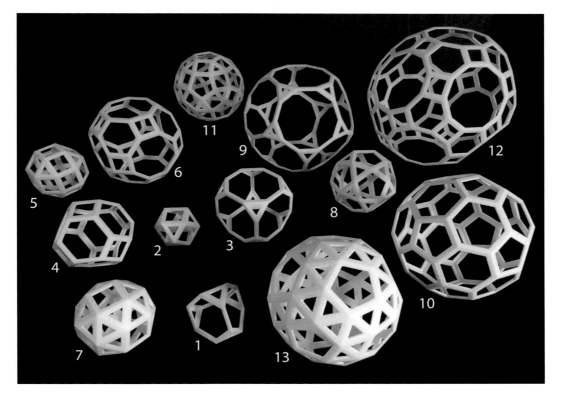

6. truncated cuboctahedron
7. snub cube
8. icosidodecahedron
9. truncated dodecahedron
10. truncated icosahedron
11. rhombicosidodecahedron
12. truncated icosidodecahedron
13. snub dodecahedron

Fig. 2.9. The archimedean polyhedra.

I won't go into where these names come from. I'll just mention that *truncated* means that the corners got chopped off. So, for example, you can get a tetrahedron back from the truncated tetrahedron by gluing the corners back on: sticking a small tetrahedron onto each of the four triangular faces of the truncated tetrahedron.

There are many wonderful patterns to be found in this list. What are the symmetry types of the archimedean polyhedra?

3 Four-Dimensional Space

What is four-dimensional space? In physics, we often think of the fourth dimension as time, but here I'm talking about four dimensions of space: four different directions to move in, all at right angles to one another.

From a mathematical point of view, we can think of four-dimensional space in much the same way as we think of two- and three-dimensional space. In two dimensions, we can specify a point in space by writing down two numbers x and y, which tell us how far to go along in two perpendicular directions from a starting point (see fig. 3.1). Here, a point is marked on the plane that is three units along in the x direction and two in the y direction. We write it as the *vector* (3,2). A two-dimensional vector is a pair of numbers. In general, we can write down a point that is x units to the right and y units above the starting point as the vector (x, y). If we need to talk about a point to the left of the starting point, then x would be a negative number, and if it is below the starting point, then y would be a negative number.

In three dimensions, we have a third perpendicular direction. To say where a point in space is, we have to add a third number, say, z. See fig. 3.2. Our

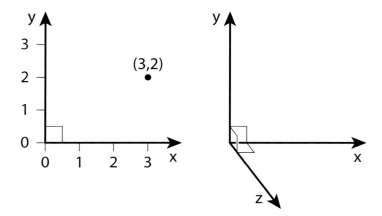

three-dimensional vectors are triples of numbers that look like (x, y, z).

In four dimensions, we (somehow) have a fourth perpendicular direction (see fig. 3.3). And we add a fourth number to say where a point is. We have run out of letters at the end of the alphabet, but we can find another one by going backward from x to w, and so our four-dimensional vectors look like (w, x, y, z).

You might want to object at this point: "Fig. 3.3 is a lie! It's impossible to find four directions that are all perpendicular to one another!" Well, it is true that the arrows drawn on the page are not all at right angles to one another, but that's *also* true of fig. 3.2. It is just as much of a lie. We are happy with fig. 3.2 only because we are three-dimensional beings, and we can interpret the figure to mean something in three-dimensional space, even though it's really just a flat, two-dimensional picture on the page. Fig. 3.2 shows a three-dimensional thing, squished down onto the two-dimensional page. In exactly the same way, fig. 3.3 shows a four-dimensional thing also squished down onto the two-dimensional page.

You might ask whether four-dimensional things are *physically real* objects. This is a (very interesting)

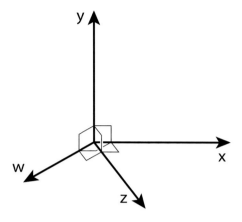

Fig. 3.3. Four perpendicular axes.

question for physics. From a mathematical point of view, however, there is no problem in talking about points in four-dimensional space using vectors of four numbers. We can measure lengths and angles in the same way as in three dimensions. Because we can imagine four-dimensional things, and we can work with them consistently, then from the mathematical point of view, they exist. The physical reality question doesn't matter so much. Four-dimensional things exist in our minds and in our calculations.

So far, we have looked at zero-, one-, two-, and three-dimensional polytopes. The only thing stopping us from going on to think about four-dimensional polytopes and higher is the problem of seeing what we are doing. We can get a sense of what is going on by squishing four-dimensional things down into three dimensions. Then our poor three-dimensional brains can try to understand these *shadows*, or *projections*, of four-dimensional objects cast onto our three-dimensional world. We have to be careful though: a shadow can give a distorted or incomplete picture of the original object. Even worse, the photographs in this book squish three-dimensional 3D prints onto the two-dimensional page, so the objects get squished twice.

Fig. 3.4. Parallel projection of a cube. The cube is sitting on a transparent sheet of plastic, which is held up above the table.

To understand the perils of squishing better, let's go back again to the situation one dimension down. Suppose that we have a two-dimensional friend who lives in a two-dimensional plane, and we want to show her something three dimensional. She can only see things in the two-dimensional plane in which she lives, so we have to show her something within her plane. One thing we could do is cast shadows of three-dimensional objects onto her two-dimensional plane, as in fig. 3.4.

Here, the light source for the shadow is far away, so the light rays are parallel (as near as makes no difference). The shadow of the cube on the table is then a *parallel projection* of the cube. The resulting shadow is not at all bad at showing our friend what the cube looks like, or at least it seems that way to our three-dimensional eyes and brains. There are some problems that she might point out though. There are places where the edges in the projection seem to cross through one another, while the true three-dimensional edges do not. The angles between some of the edges look different from one another when, again, in three-dimensional reality, they are all the same. However, there are some good points: pairs of

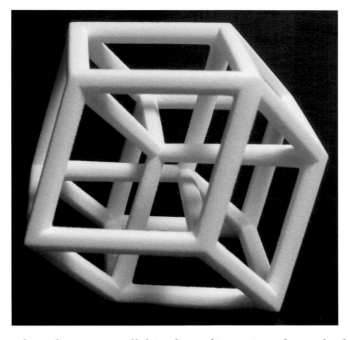

Fig. 3.5. *Hypercube B* by Bath-sheba Grossman.

edges that are parallel in three dimensions have shadows that are also parallel.

If we turn the cube around, we can get different shadows. We get a very boring square when the light is pointing directly at a face of the cube or a more interesting hexagon when the light is pointing at a corner. (See the top-right photograph in fig. 1.12.) For showing the cube to our two-dimensional friend though, it would be less confusing to shine the light in a "less special" direction, which doesn't have the shadows of vertices overlapping one another.

Moving up to four dimensions, fig. 3.5 shows a parallel projection of the four-dimensional *hypercube*, also known as the *tesseract*. This is the "shadow" on the three-dimensional "table" we live on, cast by the true four-dimensional hypercube. I can't show you a four-dimensional picture of how this 3D print is a shadow. The best I can do is to show you the result. As far as the mathematics is concerned, it really is just the same as projecting the three-dimensional cube down to two dimensions.

What is the hypercube anyway? Just as the cube is the three-dimensional analogue of the two-dimensional square, the hypercube is the four-dimensional

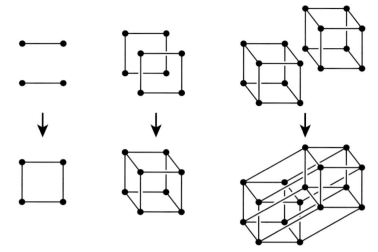

Fig. 3.6. Constructing a hyper-cube.

analogue of the cube. Fig. 3.6 shows one way to think about what it is: Start with a line segment, take a copy of it shifted to the side, connect up corresponding ends of the line segments, and you get a square. Now take a copy of the square, shifted to the side, connect up corresponding vertices of the squares, and you get a cube. Now take a copy of the cube, shifted to the side, connect up corresponding vertices of the cubes, and you get a four-dimensional hypercube.

We could continue this pattern, making the five-dimensional hypercube the six-dimensional hyper-cube and so on. Back to fig. 3.5: is there anything we have to be careful about with this 3D printed "shadow"? Does it seem to tell us something about the hypercube that isn't true of the actual four-dimensional thing? Unlike fig. 3.4, we don't have any edges crashing into each other where they shouldn't, but it does look like there are edges that crash through faces (they don't in the actual four-dimensional hypercube). However, we again get that parallel edges in the shadow are parallel in the actual four-dimensional object.

Let's explore a few other ways to see the hypercube, again going back down to projecting from three dimensions to two to help us see what's happening.

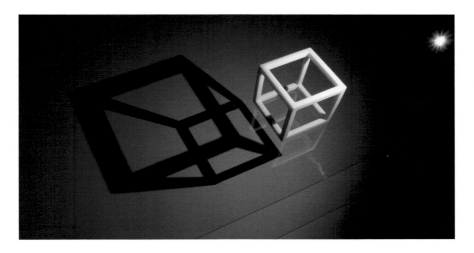

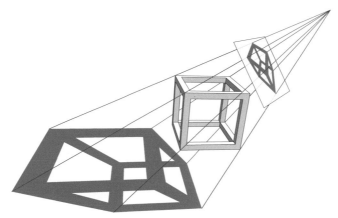

In fig. 3.7, the light source is near to the cube, and we have lost the nice feature of parallel edges always having parallel shadows. The shadow we see looks like a perspective drawing of the cube. This isn't a coincidence, as we see on the bottom of fig. 3.7. Remove the light and put your eye where the light was. Instead of light rays leaving the light and getting blocked by the cube to make the shadow, we have light rays coming from the cube into your eye and hitting your retina, making the image that you see. Rather than drawing the image on your retina, we can think of drawing the picture you would see on a little transparent screen between your eye and the cube. Now, look at the cube through the screen and trace the outline of the cube on the screen. What you see on the screen is a scaled-down version of the shadow.

Fig. 3.7. Perspective projection of a cube. *Bottom*, why the shadow looks like a perspective drawing of a cube.

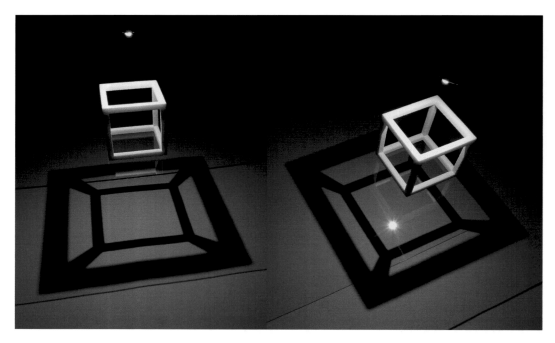

Fig. 3.8. Two views of a perspective projection of a cube with nonintersecting edges.

We can move the light around to get different shadows. If we put the light above a face of the cube, as in fig. 3.8, then the shadows of the edges don't overlap. However, the shadows of the *faces* do still overlap: there are six square faces of the cube, five of which are sent to their own regions in the plane, but the top face overlaps the other five.

Let's return to four dimensions. In fig. 3.9, no edges crash through faces. This makes it easier to see the projections of the eight cubes. There is one small cube in the center (which looks small because it is far away from the light source in the fourth dimension), six distorted cubes around the central cube, and one more big cube that overlaps all of the others.

This is almost my favorite projection, but I like one even better. Let's start again with the cube, but instead of projecting shadows onto the plane, let's first *radially* project the edges of the cube onto a sphere centered on the cube.

By putting the light at the center of the cube (see fig. 3.10), the shadow on the sphere creates a rounded "beach ball cube." The straight edges of the cube turn into arcs of circles on the sphere. This is called *radial*

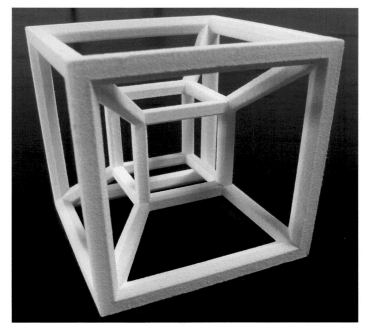

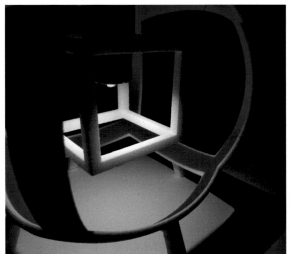

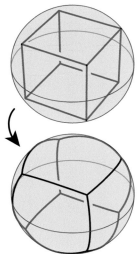

Fig. 3.9. *Hypercube A* by Bath-sheba Grossman.

Fig. 3.10. Making a beach ball cube.

projection because the light rays come out from the center of the sphere, radiating from that point.

Next, let's take this beach ball cube as a solid object and cast shadows of it onto the plane. We put the light so close that it is actually on the surface of the sphere, at its north pole. See fig. 3.11. This projection from the sphere to the plane is called *stereographic projection*, and it has some wonderful properties.

The first amazing property of stereographic project-

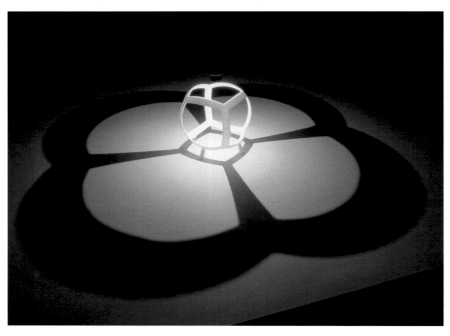

Fig. 3.11. Stereographic projection of the beach ball cube onto the plane.

ion is that it preserves angles. That is, the angle between two curves on the sphere is the same as the angle between their shadows, so the angle is preserved by the projection. The grid on the sphere in fig. 3.12 is distorted, with longer edges toward the south pole and shorter edges toward the north. However, if you look closely, you will notice that only lengths are distorted. The angles at the corners of the windows on the sphere are all right angles, as are the corresponding angles in the shadow on the table.

The curves that make up the frames of the windows on the sphere are all arcs of circles, and these all project to straight lines on the plane. These are very special arcs though. If we extend them to circles, they all go through the north pole, which is why in the projection on the plane they have to keep going outward forever. What happens with other circles on the sphere, ones that don't go through the north pole? Amazingly, these circles project to circles on the plane, as illustrated in fig. 3.13, a design with circles all over it, based on the dodecahedron and icosahedron.

Strangely, although stereographic projection knows that circles should go to circles, it gets a little confused and forgets where the centers of those circles go. You

Stereographic projection isn't directly related to stereo vision or stereo sound. The prefix "stereo" has generally come to mean that two channels of data are used, but both the prefix and the word "stereographic" come from an earlier common source. The original Greek word means *solid*. Stereographic projection is a way of showing solid objects in two dimension, while stereo vision lets you better perceive solid objects with your eyes. Being able to use stereo vision is a big advantage of looking at physical 3D printed objects, even over virtual models you can rotate around on a computer screen. Although by the time you're reading this, perhaps the book's website has turned into some sort of immersive stereo vision virtual reality experience.

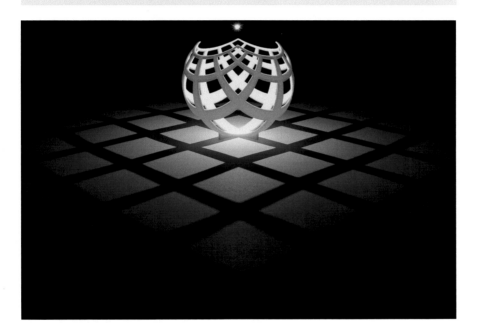

can see this on the bottom of fig. 3.13: The center of the circle on the sphere projects to the green dot. But the center of the circle on the plane (marked with the red dot) is quite a bit farther out.

Finally, notice that every point of the sphere other than the north pole is projected to some point of the plane, and every point of the plane is hit by some ray of light from the north pole.

Back to the beach ball cube and its shadow in fig. 3.11. Because stereographic projection preserves angles, the angles between the edges of the shadow are 120 degrees, as they are for the beach ball cube. The shadow looks quite similar to the shadow for the

Fig. 3.12. Stereographic projection of a grid pattern onto the plane. I made this model by starting with the grid in the plane, then using the reverse projection from the plane onto the sphere to figure out what the pattern should be.

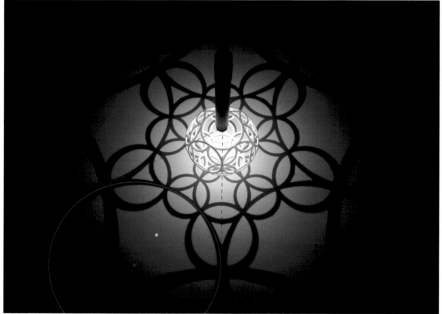

Fig. 3.13. Stereographic projection of a pattern of circles on the sphere, projecting to circles on the plane. *Bottom*, the center of the red circle is at the red dot. The center of the circle on the sphere projects to the green dot.

perspective projection in fig. 3.8, but there is another difference: now there is no overlap between the projected faces at all. In the perspective projection, the top face of the cube is projected over the other five. But in the stereographic projection, the top face of the cube is projected *outside* of the other five, covering the whole of the rest of the plane.

The projection of the cube doesn't have all of the symmetry of the cube. On the cube, all of the faces are the same, while that's obviously not true in the projection. The projection only has a single kaleidoscopic point where four mirror planes meet. Even that much symmetry only happens with a careful positioning of the cube, so that the south pole for the projection is the center of a face of the cube. This is called a *face-centered projection*. In general, the projection won't have any symmetry at all, although if we put one of the other kaleidoscopic viewpoints for the cube at the south pole, then we would get the symmetry associated to that viewpoint (and our projection would be a *vertex-centered* or an *edge-centered* projection, depending on which kaleidoscopic viewpoint we used).

As usual, let's do the same thing in four dimensions. The first thing to do is to make a beach ball hypercube by projecting the edges of the hypercube onto a sphere in four-dimensional space. Well, what is a sphere in four-dimensional space? A sphere is, by definition, the set of points all at a certain distance (the radius) from some center point. If we are working in the two-dimensional plane, then a sphere is what we would usually call a circle. The circle is the *one-dimensional sphere* (or *one-sphere* for short) because it is really a one-dimensional object—very similar to a line if you zoom in and only look at a small part of it. If we are working in three-dimensional space, then a sphere is the shape we would usually think of as a sphere (meaning the spherical surface, not the space inside). It is the *two-dimensional sphere* (or *two-sphere* for short) because it is really a two-dimensional object—very similar to the plane if you zoom in and only look at a small part of it. So, now that we are working in four-dimensional space, the sphere is the *three-dimensional sphere*, or *three-sphere* or *hypersphere*.

After we project the edges of the hypercube from its center onto the three-sphere, to make our beach ball hypercube, we can stereographically project them into three-dimensional space, and fig. 3.14 shows the result. One of the eight cubical cells of the hypercube is

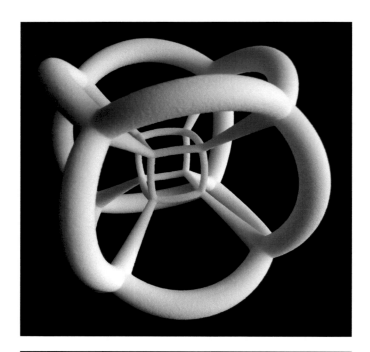

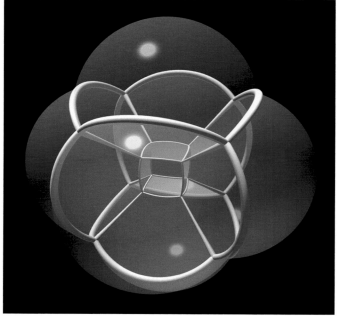

Fig. 3.14. *Top*, a 3D print of the stereographic projection of a beach ball hypercube to three-dimensional space. *Bottom*, a computer render of the same, including the two-dimensional square faces.

in the center, six more are arranged around the central cube, and the eighth is outside of the other seven. The top face of the cube in fig. 3.11 is projected to cover the whole of the rest of the plane. Likewise, the top cube of the hypercube is projected out to cover the whole of the rest of three-dimensional space. In fig. 3.14, we are *inside* the eighth cube.

Just like two-dimensional stereographic projection, three-dimensional stereographic projection preserves angles, sends circles to circles or lines, and there is no overlap between the projected faces. This is really a very accurate picture of the hypercube, or at least the beach ball hypercube.

As with the stereographic projection of the cube, stereographic projection of the hypercube again loses some symmetries. It has symmetry type *432, like the cube. However, the hypercube has even more symmetries—rotations and reflections of four-dimensional space that leave the hypercube looking the same afterward. Just as there is a symmetry of the cube that takes any flag of the cube to any other, there is a symmetry of the hypercube that takes any flag of the hypercube to any other. (Remember, this is what it means for the hypercube to be a regular polytope.) We won't go into four-dimensional symmetries in detail, but we can at least count up how many of them there are here: the same as the number of flags. A flag of the hypercube has a vertex, an edge, a square face, a cubical cell, and the hypercube itself. Every flag has the hypercube, a choice of one of the eight cubical cells, a choice of one of that cube's six square faces, a choice of one of that square's four edges, and a choice of one of that edge's two vertices. This gives us a total of $8 \times 6 \times 4 \times 2 = 384$ flags, hence, 384 symmetries.

Recall that the Schläfli symbol for the square is {4} and is {4,3} for the cube because you make a cube by arranging three squares around each vertex. The Schläfli symbol for the hypercube is {4,3,3}. The first two numbers tell us that we are making our regular polytope out of cubes, and the third number, 3, tells us that there are three cubes arranged around each edge.

Let's go back and think some more about what the three-sphere is like. See fig. 3.15. Above, we have the ordinary two-sphere cut up into eight curvy triangles, and the shadows show the result of stereographically projecting it to the plane. In fact, this is a *beach ball octahedron*—what you get when you radially project the edges of an octahedron onto the two-sphere.

Fig. 3.15. *Top*, stereographic projection of a beach ball octahedron to the plane. *Bottom,* the result of stereographic projection of a beach ball 16-cell to three-dimensional space.

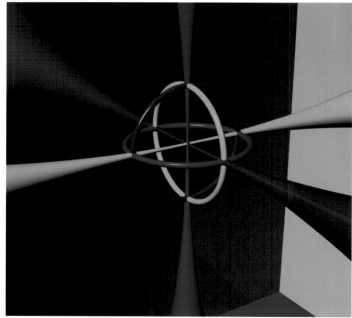

On the plane, there are four inner triangles directly below the print, coming from the four triangles in the southern hemisphere, and four more outer triangles on the outside, coming from the four triangles in the northern hemisphere. The outer triangles look strange and stretched out, because the edges of the triangles in the northern hemisphere go all the way up to the north pole. But we should think of the outer triangles and the inner triangles as the same. They only look different because of the distortion of stereographic projection. Note that every point on the two-sphere other than the north pole gets projected to a point on the plane. We can think of the two-sphere as being the plane, plus a single point added at infinity, the north pole.

Now let's look at the lower image in fig. 3.15. Here I can't show you the three-sphere sitting in four-dimensional space; I can only show you the shadow. This time, instead of four inner triangles corresponding to the southern hemisphere of the two-sphere, there are eight inner tetrahedra corresponding to the southern hemisphere (or perhaps it should be *hemihypersphere*) of the three-sphere. In the upper image, the southern and northern hemispheres of the two-sphere are separated by an equator—a circle, or one-sphere. In the lower image, the southern and northern hemispheres of the three-sphere are also separated by an equator, but this time it is a two-sphere. There are eight outer tetrahedra corresponding to the northern hemisphere of the three-sphere, and, as before, they look strange and stretched out. All 16 tetrahedra are really the same as one another, just like the eight triangles of the two-sphere in the upper image are the same as one another. The inner and outer tetrahedra only look different because of the distortion of stereographic projection. As in the two-dimensional case, apart from the north pole of the three-sphere, every point gets projected to a point of three-dimensional space. We can think of the three-sphere as ordinary three-dimensional space, with one point added at infinity, the north pole.

The upper image in fig. 3.15 is a beach ball octahedron and its projection to the plane. The lower image is the projection of another beach ball polytope, a four-dimensional polytope called the *16-cell*. As with the polygons and polyhedra, the four-dimensional polytopes (or *polychora*, singular: *polychoron*) are named for the number of "sides" they have. The sides

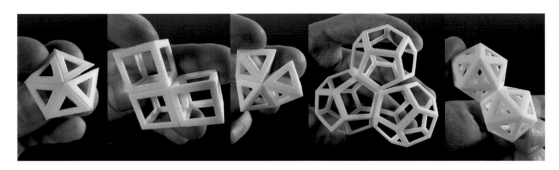

Fig. 3.16. Polyhedra fitting around an edge. Five tetrahedra, three cubes, three octahedra, and three dodecahedra, but only two icosahedra can fit.

here are three-dimensional polyhedra, called *cells*, as in the cells of a honeycomb. The 16-cell has 16 tetrahedral cells, hence the name.

We have mentioned two regular, four-dimensional polytopes, the hypercube (or 8-cell) and the 16-cell. What are the possibilities for the other regular four-dimensional polytopes? This is a similar problem to the three-dimensional case, except that instead of arranging regular polygons around a vertex (as in fig. 2.7), we have to arrange regular polyhedra around an edge. See fig. 3.16.

In chapter 2, we built a cube, starting with arranging three squares around a vertex in the plane. Because the angles at the vertex add up to 270 degrees, which is less than 360 degrees, we can fold two of the squares up into the third dimension so that the three squares form a corner of a cube. Similarly, we can arrange three cubes around an edge in three-dimensional space. Because the angles at the edge (the *dihedral angles* of the cubes) add up 270 degrees, which is less than 360 degrees, we can fold two of the cubes up into the fourth dimension so that the three cubes form an edge of a hypercube. (Of course, this isn't easy to visualize.) We can figure out what the possibilities are

Table 3.1 The regular polychora

Schläfli symbol	Vertices	Edges	Faces	Cells	Name
{3,3,3}	5	10	10	5	Pentachoron, or 5-cell
{4,3,3}	16	32	24	8	Octachoron, or hypercube or tesseract or 8-cell
{3,3,4}	8	24	32	16	Hexadecachoron, or 16-cell
{3,4,3}	24	96	96	24	Icositetrachoron, or 24-cell
{5,3,3}	600	1,200	720	120	Hecatonicosachoron, or 120-cell
{3,3,5}	120	720	1,200	600	Hexacosichoron, or 600-cell

by looking at the dihedral angles of the regular polyhedra to see how many can fit around an edge before the angle gets to or larger than 360 degrees.

Four cubes doesn't work because the dihedral angles add up to 360 degrees, which would be flat. Up to five regular tetrahedra can fit around an edge, so three, four, or five tetrahedra around an edge are possibilities, corresponding to Schläfli symbols {3,3,3}, {3,3,4}, and {3,3,5}. Three octahedra can fit around an edge, giving {3,4,3}, but four cannot. Three dodecahedra can fit around an edge, giving {5,3,3}, but four cannot. See table 3.1. Finally, three icosahedra cannot fit around an edge. All six of these possibilities in fact correspond to regular four-dimensional polytopes.

Let's see what these polychora look like, starting with the 5-cell. As with the hypercube, we can visualize this by first radially projecting onto the three-sphere, then stereographically projecting to three-dimensional space. Fig. 3.17 shows what we get. As the name suggests, five tetrahedral cells form the five "sides." In fig. 3.17, four of the tetrahedra are arranged around the central vertex, and, as in fig. 3.14, we are inside of the last cell. We can also count the number of vertices, edges, and faces.

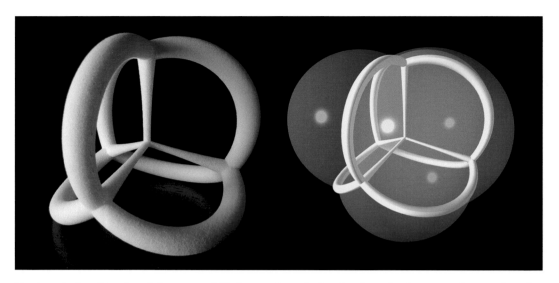

Fig. 3.17. *Left*, a 3D print of the stereographic projection of a beach ball 5-cell to three-dimensional space. *Right*, a computer render of the same, including the two-dimensional triangular faces.

We have already looked at the hypercube in detail; let's move on to the 16-cell. As the objects become more complicated, it gets harder to understand the shape from single photographs. Instead of a single photograph, fig. 3.18 shows a grid of views I made using the same rig I used in chapter 1. The model here is the *cell-centered projection* of the 16-cell, meaning that the 16-cell is arranged so that the south pole of the three-sphere is at the center of one of the tetrahedral cells. The lower image in fig. 3.15 shows the *vertex-centered projection*, so called because there is a vertex in the middle. The opposite vertex is at the north pole of the three-sphere before we project it. That vertex goes "to infinity" in three-dimensional space. This is why the background of the picture is colored red, green, and blue: these are where the red edge to the left, the green edge to the right, and the blue edge below meet near the vertex at infinity—that is, very far away.

The cell-centered projection of the 16-cell in fig. 3.18 has three-dimensional symmetry type *332. The figure shows views spaced evenly over the panel of possible views for the *332 symmetry type (see fig. 1.29).

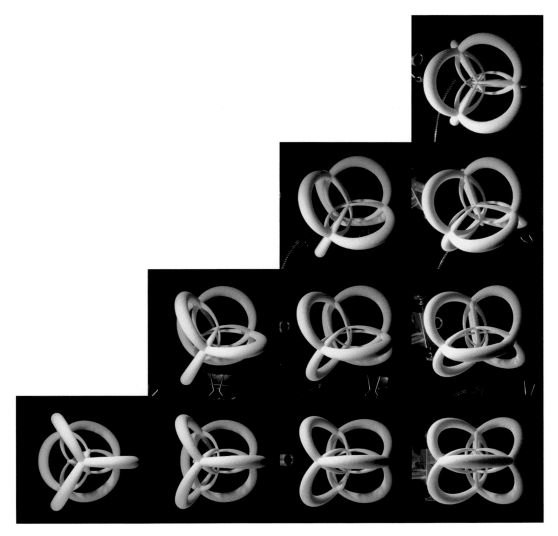

Next, fig. 3.19 shows views of the cell-centered projection of the 24-cell in the panel of possible views for the *432 symmetry type. This is already a complicated object, and we haven't even got to the 120- and 600-cells yet. We can simplify the picture by only printing part of the object. Let's look at the part of the 24-cell in the southern hemisphere of the three-sphere—what you get by cutting the three-sphere along the equatorial two-sphere halfway between the north and south poles, and throwing away the northern hemisphere. See fig. 3.20. Rings of edges of the 24-cell lie on the equatorial two-sphere, which are all cut in half when we take only the southern hemisphere.

Fig. 3.18. A grid of views of the 16-cell.

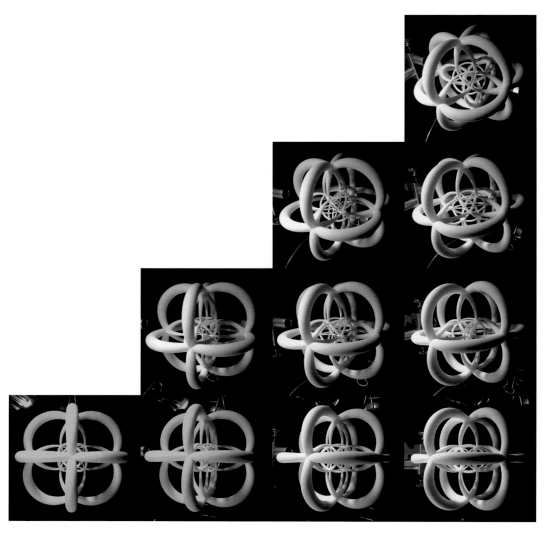

Fig. 3.19. A grid of views of the 24-cell.

We can get back the whole of the 24-cell by reflecting across the equator. This half-24-cell isn't so hard to understand. You can see that there's one octahedron in the center, with another eight octahedra attached to its eight faces. Then there are six more half-octahedra, each cut off by the equatorial two-sphere, each touching the central octahedron at one of its corners. Let's check: $1 + 8 + 6 \times 1/2 = 12$, which is half of the total 24 cells.

This trick of only looking at half of the polytope helps again when we look at the 120- and 600-cells. Fig. 3.21 shows a computer render of the whole of the (cell-centered projection of the) 120-cell. This time

Fig. 3.20. *Top,* Half of the 24-cell.

Fig. 3.21. *Bottom,* The 120-cell.

the cell-centered projection has symmetry type *532. This is a very complicated shape, with lots of hard-to-see internal structure. It's a bit easier to under-stand half of it, as seen in fig. 3.22. This is still really complicated though, so complicated that my other trick of showing you a grid of views from different directions doesn't really help any more. Instead, I've only put a few bigger pictures in the figure. We are bumping up against what is possible to see from two-dimensional photographs alone. To get a sense

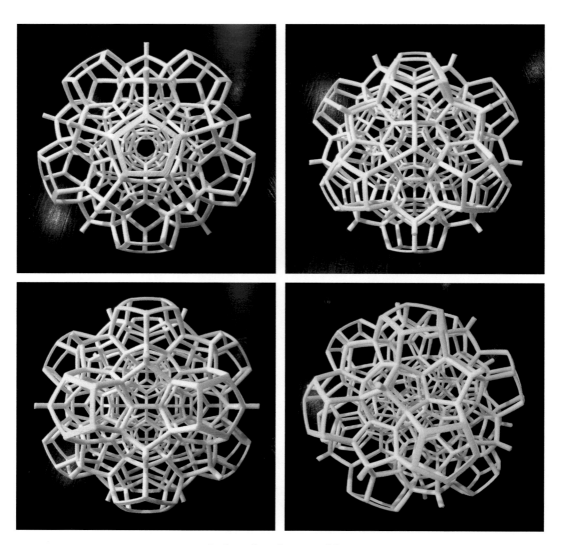

Fig. 3.22. Half of the 120-cell.

of what the shape is like, I encourage you to look at a real-world 3D print or the virtual 3D model on the website (3dprintmath.com). Finally, fig. 3.23 shows half of the vertex-centered projection of the 600-cell, which also has symmetry type *532.

The 24-cell may not be so hard to comprehend, particularly if we only look at half of it, but even half of the 120- and 600-cells seem to be amazingly complex objects. We need some new tricks to understand them. The first new trick is that working out something for one of them automatically tells us something about the other, by *duality*, which we will investigate next.

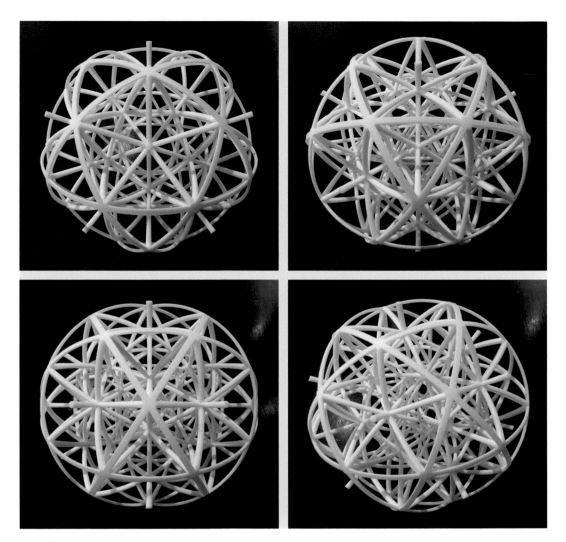

Duality

You may have noticed a striking pattern in table 3.1, in the numbers of vertices, edges, faces, and cells of the different polychora. For some of these polychora, the 5-cell and the 24-cell, there are the same number of vertices as cells, and the same number of edges as faces. The list of numbers is palindromic; it looks the same when you read it backward. The other four regular polychora come in pairs: if you take the list of numbers for the hypercube and read it backward, you get the list of numbers for the 16-cell and similarly for the 120-cell and the 600-cell.

Let's look back at the three-dimensional polyhedra.

Fig. 3.23. Half of the 600-cell.

Table 3.2 The regular polyhedra				
Schläfli symbol	Vertices	Edges	Faces	Name
{3,3}	4	6	4	Tetrahedron
{4,3}	8	12	6	Hexahedron, or cube
{3,4}	6	12	8	Octahedron
{5,3}	20	30	12	Dodecahedron
{3,5}	12	30	20	Icosahedron

Fig. 3.24. Dual polyhedra. *From left to right*: the tetrahedron is self-dual, the cube and octahedron are duals of each other, and the dodecahedron and icosahedron are also duals of each other.

See table 3.2. Again, we see the same pattern. The tetrahedron is palindromic, the cube is paired with the octahedron, and the dodecahedron is paired with the icosahedron. This phenomenon is called *duality*. The cube and octahedron are duals of each other, as are the dodecahedron and the icosahedron, while the tetrahedron is *self-dual*. Similarly, the hypercube and 16-cell are duals of each other, as are the 120-cell and 600-cell, while the 5-cell and the 24-cell are each self-dual.

There is a neat way to figure out which polyhedra and polychora are dual to each other. If you reverse the numbers in the Schläfli symbol for a polytope, you get the Schläfli symbol for its dual.

Let's investigate the cube and the octahedron first. Look at the middle 3D print in fig. 3.24. Start with the cube, which we will think of as colored blue, and radially project it to the sphere to make a blue beach ball cube. Now we can build a red beach ball octahedron as follows: put a red vertex at the center of each of the faces of the beach ball cube. Whenever two of the cube's blue faces meet at a blue edge, draw a red edge between the red vertices at the centers of the blue faces. These red vertices and edges cut the sphere up into triangles, with a blue vertex in the middle of each. This makes the beach ball octahedron.

Because one of the cube's vertices is in the middle of each of the octahedron's faces, there are the same number of each. Similarly, there are the same number of vertices of the octahedron as there are faces of the cube. We can also see that there are the same number of edges in each, because each of the cube's edges crosses one of the octahedron's edges and vice versa.

The same process also works to make the cube from the octahedron, the icosahedron from the dodecahedron, and vice versa. If we apply it to the tetrahedron, we get another tetrahedron back.

We can build the dual of a four-dimensional polychoron in much the same way. See fig. 3.25. Again, we use radial projection to make beach ball versions of the polychora, this time on the three-sphere rather than on the two-sphere. Stereographic projection lets us see the three-sphere as three-dimensional space (plus a point at infinity), and all of these are prints of projections of the southern hemisphere of the three-sphere. This time we start with a (blue, say) polychoron and build its (red) dual as follows. For each blue cell, put a red vertex at its center. When two blue cells meet at a blue face, draw a red edge between the corresponding red vertices. Then each blue edge goes through a red face that we can draw in, and these red faces cut the three-sphere into red cells, each of which has a blue vertex at its center. Note that the half 16-cell shown is the vertex-centered projection, which is dual

Fig. 3.25. Dual polychora. *Top left*, the 5-cell is self-dual; *center*, the hypercube and 16-cell are duals of each other; *bottom left*, the 24-cell is self-dual; *bottom right*, the 120-cell and 600-cell are duals of each other.

to the cell-centered half hypercube. Similarly, we see both vertex-centered and cell-centered half 5-cells and half 24-cells.

Before we move on from duality, there are many, many other beautiful relations among the regular polytopes. One of the prettiest: it turns out that the 20 vertices of the dodecahedron can be divided into five groups of four vertices, each group being the vertices of a regular tetrahedron. George Hart's 3D print illustrates this fact (see fig. 3.26). What is the symmetry type of this object (ignoring the colors)?

The 120-Cell

Let's return to the 120- and 600-cells. Because they are dual to each other, if we understand how one of them is put together, we also know how the other is put together, so we can just concentrate on the 120-cell. One way to understand the 120-cell is to look at layers of dodecahedra at different distances from the south pole. See fig. 3.27. We start at the far left, with one red dodecahedron centered on the south pole. This dodecahedron is the cell that makes this the cell-centered projection of the 120-cell. Next to the red dodecahedron are 12 orange dodecahedra.

Fig. 3.26. *Five Tetrahedra Centerpiece*, by George W. Hart. Five tetrahedra fit inside a dodecahedron (not shown). Each tetrahedron has vertices at four of the twenty vertices of the dodecahedron.

Around them are 20 yellow dodecahedra and then another layer of 12 green dodecahedra. Next is a layer of 30 blue dodecahedra. This layer is on the equatorial two-sphere, halfway to the north pole. The circle on the right of fig. 3.27 shows schematically where these layers are in the three-sphere relative to the light at the north pole.

After the equatorial layer, four more layers (not shown) repeat the pattern of the first four in reverse order all the way to the final dodecahedron centered on the north pole. Let's count dodecahedra to make sure we haven't missed any. Up until the equator, we have 1 + 12 + 20 + 12 = 45 dodecahedra in the southern hemisphere. Then we have another 30 on the equator and an additional 45 in the northern hemisphere. In total, we have 45 + 30 + 45 = 120 dodecahedra.

Fig. 3.28 shows a different way to break the 120-cell up into easier-to-understand pieces. In the top left, is a gray ring of 10 dodecahedra on the equatorial layer. To make the ring, imagine starting inside of a dodecahedron and traveling through one of its pentagonal faces into a neighboring dodecahedron. In this new dodecahedron, opposite the pentagonal face we entered through is another pentagonal face. We travel

  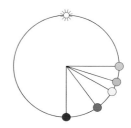

Fig. 3.27. The layers of dodecahedra around the south pole, going out as far as the layer on the equator. *Right,* a schematic diagram of the positions of the layers in the three-sphere.

straight through the dodecahedron, out through this opposite pentagonal face, and carry on going. Eventually, we come back to the first dodecahedron having visited 10 dodecahedra.

If we start with a new dodecahedron next to our first gray ring, and travel alongside it, we make another ring of 10 dodecahedra that wraps around the first. Five rings nestle around the first one for a total of six rings of 10 dodecahedra, or 60 dodecahedra. This is half of the 120-cell. Another five rings (not shown) wrap around these, and then there is one more ring (also not shown) that goes vertically down through the hole in the middle to complete the 120-cell.

Saul Schleimer is a mathematician at the University of Warwick. He and I often work on projects together. We designed a 3D printed illustration of this arrangement of rings (in fact, only the central ring and two of the rings that wrap around it). Fig. 3.29 shows how it works. The three loose rings fit snugly together to give part of the cell-centered projection of the 120-cell in two ways shown in the figure, and one further very surprising way (Saul and I were not expecting it). The two ways shown in fig. 3.29 have the rings fitting into the "wrapping around" scheme, but the third

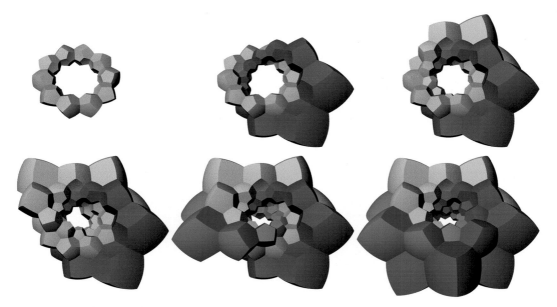

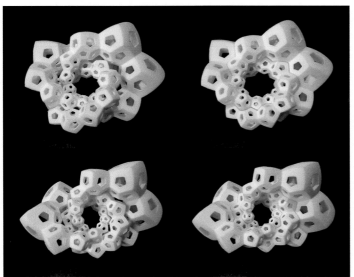

Fig. 3.28. *Top,* Rings of dodeca-hedra.

Fig. 3.29. *Bottom,* 3D printed rings of dodecahedra from the cell-centered projection of the 120-cell. The top two images show one way in which these rings go together: on the left just before they fit into place, and on the right after. The bottom two images show a second way in which the rings fit together.

unexpected way does not. I'll leave as a (very difficult) puzzle for you to ponder what that third way is.

Saul and I took this investigation a step further, to make a family of construction puzzles based around fitting parts of rings of dodecahedra together snugly as cells of the cell-centered projection of the 120-cell. See fig. 3.30. The different "rib" pieces shown in the top left can fit together into a bewildering array of different constructions.

We call the family of puzzles in fig. 3.30 "Quintessence" after Plato, who thought that the cosmos itself was made from dodecahedra, just like our puzzles. Plato associated the four classical elements, fire, earth, air, and water, with the regular polyhedra: Earth is supposedly made out of cubes because you can stack them together to make a solid object, while fire is made from tetrahedra, because they are sharp and pointy, and hurt when you step on them. Water is made from icosahedra because they roll and flow so well, and somewhat more tenuously, air is made out of octahedra because they are so smooth that one can barely feel them. The fact that there are five regular polyhedra but only four classical elements clearly wasn't enough to keep a good idea down, hence the addition of the fifth element, the quintessence.

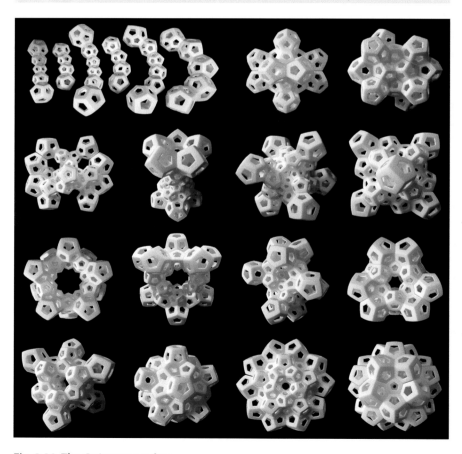

Fig. 3.30. The *Quintessence* family of puzzles, by Saul Schleimer and Henry Segerman.

More Fun Than a Hypercube of Monkeys

We briefly looked at four-dimensional symmetries earlier in this chapter, calculating that the hypercube has 384 symmetries. A similar calculation shows that the 120-cell has 14,400 symmetries. As you might imagine, the classification of symmetries is considerably more complicated in four dimensions. See *On Quaternions and Octonions: Their Geometry, Arithmetic, and Symmetry* by John H. Conway and Derek A. Smith if you want to see the full, gory details.

One of the simplest and most beautiful four-dimensional symmetry types is given by something called the *eight-element quaternion group*. There are eight symmetries, meaning that there are eight motions of the object in four-dimensional space that leave it looking the same. Fig. 3.31 shows two views of the stereographic projection of a sculpture in four-dimensional space with this symmetry type. Unlike the stereographic projections of the polychora, this object loses all of its symmetry when projected. However, we can still see the symmetry in this distorted form. There are eight monkeys, all of which are identical. Each monkey sits inside one of the eight cubical cells of the hypercube and has a limb, a tail, or a head connected to a limb, a tail, or a head of a monkey in each of its six neighboring cubes. Each monkey body is at a vertex of the 16-cell dual to the hypercube, and each monkey limb/tail/head connection corresponds to an edge of the dual 16-cell. The projection is cell centered relative to the dual 16-cell, so the projected sculpture looks similar to the cell-centered projection of the 16-cell shown in fig. 3.18. There are two rings of four monkeys connected by the head/foot connection, two rings of four monkeys connected by the hand/tail connection, and two rings of four monkeys connected by the hand/foot connection. These two rings of four monkeys are precisely analogous to the 12 rings of 10 dodecahedra in the 120-cell.

Fig. 3.32 shows one of the monkeys inside its cube together with the connections to its six neighboring monkeys, before they are folded up to form the hy-

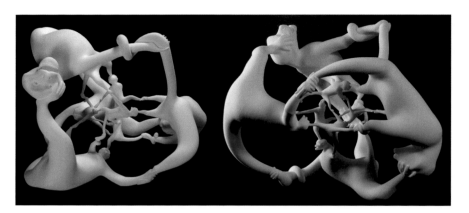

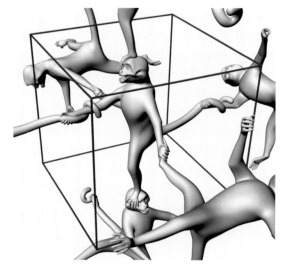

Fig. 3.31. *Top,* Two views of *More fun than a hypercube of monkeys*, by Will Segerman and Henry Segerman, inspired by a question of Vi Hart.

Fig. 3.32. *Bottom,* A monkey in a cube, with its neighbors.

percube. The eight symmetries take any one monkey to each of the eight monkeys. The do-nothing symmetry takes a monkey to itself. Six of the seven other symmetries are screw motions, in which we rotate by a quarter turn to the left while also rotating along one of the rings of monkeys. Finally, the eighth symmetry is what we get by doing any screw motion twice. It takes any monkey to the monkey in the cube on the opposite side of the hypercube. For more details on what is going on here, see my article with Vi Hart, *The*

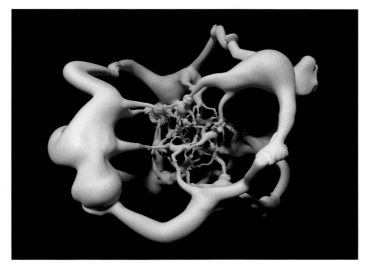

Fig. 3.33. *More fun than a 24-cell of monkeys*, **by Will Segerman and Henry Segerman.**

Quaternion Group as a Symmetry Group (details in appendix A, fig. 3.32). You can see an animated version of the sculpture rotating in four dimensions online at http://monkeys.hypernom.com.

Why stop with the hypercube? The 24-cell and the 120-cell have their own screw motion, four-dimensional symmetry types, in the same way that the hypercube has the eight-element quaternion group. So, of course, we also made monkey sculptures with those symmetries. Fig. 3.33 shows what you get for the 24-cell: 24 monkeys form four rings of six monkeys each. Fig. 3.34 shows the 120-cell version. Because it would be enormous if we printed all 120 monkeys, the sculpture has 45 monkeys in the southern hemisphere together with 30 equatorial monkeys, but we left out the final 45 monkeys in the northern hemisphere. So 11 of the 12 head/foot rings of 10 monkeys are cut off when they go into the northern hemisphere. The one other complete ring is made entirely out of equatorial monkeys.

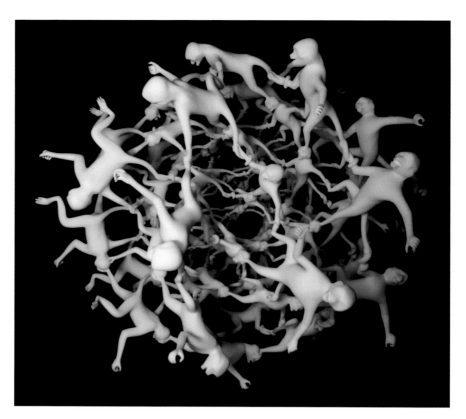

Fig. 3.34. *More fun than a 120-cell of monkeys*, by Will Segerman and Henry Segerman.

4 Tilings and Curvature

Let's return to two dimensions, and the question of zero, or negative, angle defect. We already thought a little about this when we tried to make a polyhedron using six equilateral triangles around each vertex, which would have Schläfli symbol {3,6} and zero angle defect at each vertex.

We can build this, not as a polyhedron, but as a tiling of the plane. The same is true for {4,4}, the tiling of the plane with four squares around each vertex, and {6,3}, the tiling of the plane with three hexagons around each vertex. See fig. 4.1.

As we did in chapter 3, rather than thinking of straight-sided polyhedra, we can think of radially projected beach ball versions of the faces, edges, and vertices of the polyhedra. These are really polygonal tilings of the sphere. Before radially projecting, some of the polyhedra are geometrically closer to the sphere than others: the tetrahedron is very spiky (Plato would agree), whereas the icosahedron is much closer to being spherical. But after we radially project onto the sphere, all of the polyhedra become nice, regular tilings of the sphere.

So far, then, all of the (length-two) Schläfli symbols are tilings of either the sphere (if the angle defect

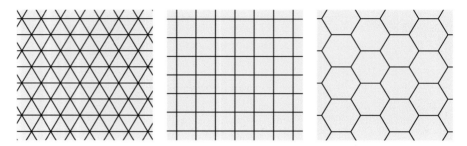

Fig. 4.1. Regular tilings of the plane.

is positive) or the plane (if the angle defect is zero). What about negative angle defect?

The thing to do is try to make a polyhedron using seven equilateral triangles around each vertex and see what we get. Then the Schläfli symbol would be {3,7}, and the angle defect would be 360 – (7 × 60) = –60. See fig. 4.2.

This doesn't close around to form an orderly pattern on a sphere, nor does it lie flat to make an orderly pattern on the plane. If we try to flatten it out in one place, it bunches up in another. In fact, it doesn't seem to make any orderly pattern at all. This is very different from the regular polyhedra, whose vertices all lie neatly spaced on a sphere, and the tilings of the plane, whose vertices lie neatly spaced on the plane. Is there a smooth surface that this tiling fits onto, like the sphere or the plane?

We can start thinking about this by trying to remove the bunching up. The problem is that there are too many triangles to fit on a flat surface. It has to ripple in order to fit into space, like a leaf of lettuce. It also seems happier in some sort of saddle shape, like a mountain pass between peaks, as in the right-hand photograph in fig. 4.2. Admittedly, it's a little hard to

see from the picture how it resembles a horse's saddle. Looking ahead a bit, the right-hand surface in fig. 4.3 is a clearer picture of a saddle. As you move away from the middle (where you sit on the saddle), the surface curves both down (where your legs go) and up (where the horse goes). Or maybe it would be better to say that the surface looks like a Pringles crisp.

Anyway, unlike our {3,7} tiling, the tilings of the plane are flat, and the regular polyhedra are shaped like the top of a hill (or a sphere). These are differences in the *curvature* (specifically, *gaussian curvature*) of the surfaces, so we will need to understand something about this first before we can understand our {3,7} tiling.

Fig. 4.2. A tiling in which seven equilateral triangles meet at each vertex. This is printed with hinges so that it can be arranged in many different ways—there doesn't seem to be one best way to arrange it in space.

Curvature

Gaussian curvature, named after Carl Friedrich Gauss, measures the extent to which a surface is not flat. The curvature of the surface near a point measures how much it looks like a hill or a bowl (*positive* curvature), a flat plane (*zero* curvature), or a saddle (*negative* curvature).

Let's choose a point on a surface and draw a small circle around it (say, with radius *r*). A circle here

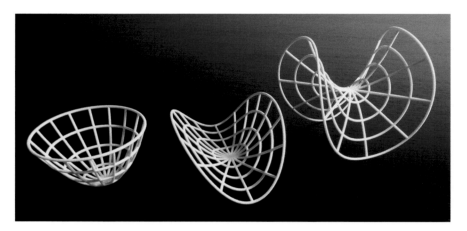

Fig. 4.3. Circles drawn around a point in surfaces with positive, zero, and negative curvature. These surfaces are the circular paraboloid (*left*), the parabolic cylinder (*middle*), and the hyperbolic paraboloid (*right*). Or rather, these are patches of these surfaces. They can be extended outward forever. On all three surfaces, the radial spokes are shortest paths along the surface from the center out to other points.

means the set of points that are at distance r from the center, where we measure the distance along the surface. Zero curvature corresponds to the circumference of this circle being $2\pi r$, as you would expect for a circle in the plane. Positive curvature corresponds to the circumference of the circle being smaller than $2\pi r$, and negative curvature corresponds to the circumference of the circle being bigger than $2\pi r$.

Fig. 4.3 shows surfaces with each of these three possible kinds of shape, the *circular paraboloid* on the left, the *parabolic cylinder* in the middle, and the *hyperbolic paraboloid* on the right. All three surfaces are patterned with equally spaced concentric circles and radial spokes.

For the bowl-like circular paraboloid on the left of fig. 4.3, the circles are too short, corresponding to positive curvature. For the hyperbolic paraboloid (on the right), the circles are too long, corresponding to negative curvature. Perhaps surprisingly, the parabolic cylinder (in the middle) has zero gaussian curvature even though it is curved. To see this, imagine unbending the surface, flattening it out onto the plane. The surface doesn't get stretched at all when you do this, so any circles drawn on it are just ordinary circles on the plane, bent onto the surface. This means that their circumferences are all $2\pi r$, as they should be for circles in the plane. Going the other direction, if you take a flat sheet of paper or fabric and roll it into a cylinder or cone, the gaussian curvature is still zero. When making clothes out of fabric this is a problem because clothes are supposed to fit on people, and people mostly don't have zero gaussian curvature.

Gaussian curvature isn't about convexity versus concavity. If you turn the circular paraboloid over, it will switch from appearing concave (like a bowl) to appearing convex (like a hill), but the curvature won't change. Looking into the bowl side, the surface bends toward you in all directions. Looking at the hill side, the surface bends away from you in all directions. However, the hyperbolic paraboloid bends away from you in some directions and toward you in other directions, and this neither-convex-nor-concave property is what gives a surface negative curvature.

A surface can have different curvatures in different places. It could be very hill-like in some places but flatter in others or even be saddle shaped in some places but hill-shaped in others. See fig. 4.4. Here are the circular paraboloid and the hyperbolic paraboloid again, shaded according to the curvature at each point. Points at which the surface is more positively curved are colored red, more negatively curved colored blue, and zero curvature colored white. The most positively curved point of the circular paraboloid is in the center, and the curvature decreases as we move away from the center, as the surface becomes flatter. The same thing happens with the hyperbolic paraboloid, except with negative curvature rather than with positive.

Also shown in fig. 4.4 are the sphere and the *torus*, the surface of a doughnut. The torus has negative curvature on the inside of the hole and slightly positive curvature around the outside. There is a circle of zero curvature at the very top of the torus and another on the bottom. A sphere has the same positive curvature

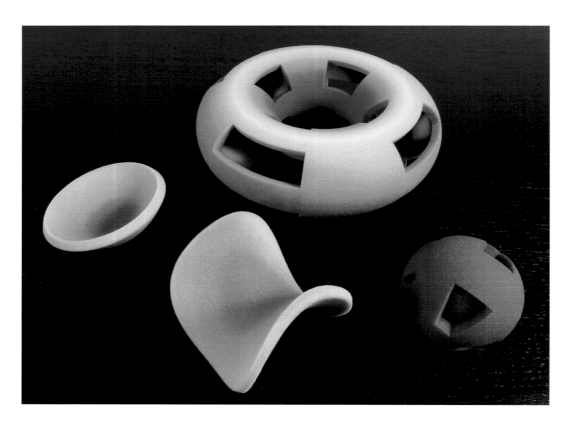

Fig. 4.4. *Counterclockwise from left*: circular paraboloid, hyperbolic paraboloid, sphere, and torus, colored according to the curvature at each point of the surface. Here red corresponds to positive curvature, white corresponds to zero curvature, and blue corresponds to negative curvature.

everywhere. This shouldn't be surprising given the sphere's symmetry.

All of these 3D prints are colored according to the same curvature scheme. For example, the center of the circular paraboloid has the same shade of red as the sphere, and they have the same curvature at that point. You can even put the sphere inside of the circular paraboloid and see how they match up. Similarly, the center of the hyperbolic paraboloid has the same curvature as the inner circle of the torus. However, they don't fit perfectly together. Curvature doesn't determine everything about the shapes of the surfaces.

The sphere and torus prints are hollow here, with the inner surface colored the same as the outer. The inner surface of the sphere curves slightly more tightly than the outer surface of the sphere, so if we were coloring its curvature separately, it should be a slightly deeper red. In general, the smaller the radius of a sphere, the larger the curvature. A sphere with a very large radius looks very much like a flat plane; therefore, it has very small positive curvature everywhere.

Back to our polygonal surfaces. It looks as though the {3,7} surface should have negative curvature, because it seems happier in a saddle shape than trying to lay it flat. But there is a subtlety here. To be clear, the model shown in fig. 4.2 has hinges and holes in the triangles, and the hinges can wiggle a bit, but let's pretend that the surface is made from solid triangles hinged perfectly along their edges. Curvature is measured at a point of a surface and can be different at different points. Suppose we choose a point in the middle of a triangle and draw a little circle around it, small enough to be contained in the triangle. Then it will have circumference $2\pi r$, because the triangle is flat. So the curvature at this point is zero. The same is true for points on the edges: just like when we flattened out the parabolic cylinder in fig. 4.3, we can flatten out any angle at the edge, so that the circle around the point is drawn in the plane, and again the curvature is zero. The story is different at a vertex though. If we draw a small circle around a vertex then it goes through seven equilateral triangles. Each equilateral triangle contributes $(1/3)\pi r$ to the circumference for a total of $(7/3)\pi r$, which is bigger than $2\pi r$ by $(1/3)\pi r$. So the surface does indeed have negative curvature, but it is all concentrated at the vertices.

This is the same situation as with an icosahedron. See fig. 2.6. It is a polygonal approximation to a sphere, and although the curvature inside each triangle and on the edges is zero, we get positive curvature concentrated at the vertices. In fact, for polygonal surfaces, the curvature at a vertex is really the same as the angle defect from chapter 2. I'll come back to this connection in chapter 6.

The icosahedron is a good approximation to a sphere, but we can make an even better approximation, with less angle defect at each vertex, by subdividing its triangles as in fig. 4.5. Fig. 4.6 shows how this works: The first model (*left*) shows five triangles

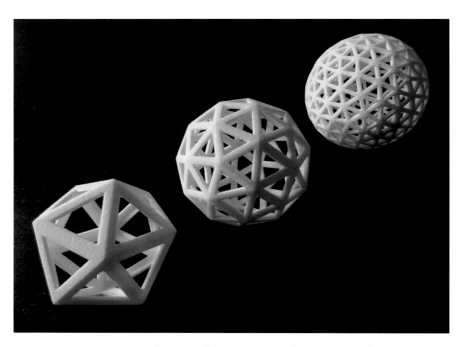

Fig. 4.5. *Bottom left to top right*, an icosahedron, a polyhedron made by subdividing the icosahedron's triangles, and one further subdivision.

arranged around a vertex, as they are for the icosahedron. The second model (*middle*) is obtained from the first by subdividing each triangle into four smaller triangles. One of these quadruples of triangles is highlighted in red in fig. 4.6. This introduces new vertices at the midpoints of the original edges, which we push out to be the same distance from the center of the icosahedron as the original vertices. In other words, the new vertices lie on the sphere that circumscribes the icosahedron. The third model (*right*) is obtained from the second by the same process, and we could keep going forever, making better and better polygonal approximations to the sphere. These subdivided icosahedra, and polyhedra obtained by similar processes, are called *geodesic spheres*, as named and popularized by R. Buckminster Fuller.

By subdividing an icosahedron (the {3,5} tiling), we can make a better approximation to a sphere, with less angle defect at each vertex. If we do the same thing to the {3,6} tiling of the plane by equilateral triangles, we just get a scaled-down version of the same tiling. See fig. 4.7. The angle defect at each vertex is zero, and it remains zero after subdividing triangles.

However, subdividing the triangles of the {3,7}

Parts of geodesic spheres are often used as architectural *geodesic domes*—a portion of a geodesic sphere makes a very lightweight and strong structure that can be made from mass-produced straight building materials (for the edges of the polyhedron). Having said that, people are already working on 3D printing entire buildings. When the technology improves to the point that 3D printed building materials are just as strong as conventionally produced materials, there would be no need to make a polygonal approximation to a sphere. We could just print a sphere.

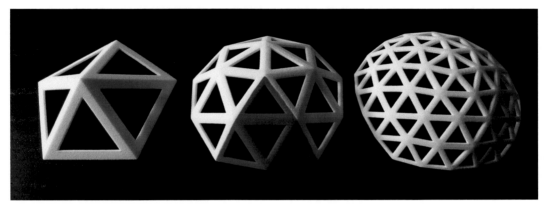

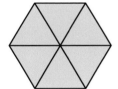 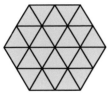 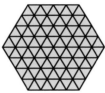

Fig. 4.6. *Above,* Geodesic domes. Each dome is obtained from the previous by subdividing the triangles into four smaller triangles. One of these quadruples of triangles is highlighted in red on the second dome.

Fig. 4.7. *Left,* Subdividing the tiling of the plane by equilateral triangles gives a scaled-down version of the same tiling.

tiling can make the angle defect at the vertices closer to zero, which should make a better approximation to whatever it might be an approximation of. See fig. 4.8. The four highlighted blue triangles came from one of the original triangles in the {3,7} tiling. It's a little difficult to see, but there are two different kinds of triangles in fig. 4.8. The middle triangle of the four blue triangles is equilateral, while the three around it are slightly isosceles, with the side touching the middle triangle being slightly shorter than the other two. In the second print shown in fig. 4.6, it is the other way round: the middle triangle is again equilateral and the others are isosceles, but with the side touching the middle triangle being slightly longer. In the positive curvature world of the sphere, when a triangle is

Fig. 4.8. A hinged surface made by subdividing each triangle of a {3,7} tiling into four triangles. Four triangles that came from subdividing one of the triangles of the {3,7} tiling are highlighted in blue.

subdivided the result bulges out. That is to say, the new triangle we get by looking at the four triangles it was subdivided into is a little fatter. However, in the negative curvature world, the subdivided triangle is a little skinnier than it was before.

This subdivided version of the {3,7} tiling feels much smoother than the {3,7} tiling itself. It hangs like fabric. Large patches of it can be laid flat on a table, although there is still some inevitable bunching up. And it still isn't at all obvious what kind of surface it might fit onto in a nice regular way.

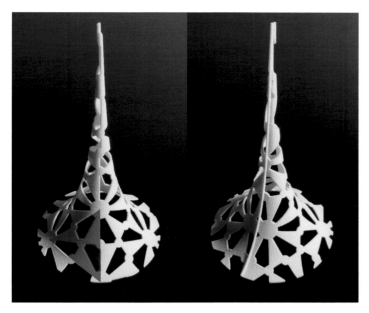

Fig. 4.9. The pseudosphere. Shaded in blue is an equilateral triangle of the same size as the triangle in fig. 4.8. *Right*, the pattern does not match up nicely around the back of the pseudosphere.

Well, it's time to give the game away about what's going on here. We have been skirting around the issue of finding a smooth surface whose gaussian curvature is negative and is the same for every point of the surface. The hyperbolic parabola in fig. 4.4 doesn't work: the curvature is negative everywhere, but it gets closer to zero as we move out from the center. If we want a surface with the same positive curvature everywhere, then we have the sphere, and if we want a surface with zero curvature everywhere, then that's the plane. And all of the tilings {3,3}, {3,4}, and {3,5} (the tetrahedron, octahedron, and icosahedron) fit nicely onto the sphere, while the tiling {3,6} fits nicely onto the plane. So we would expect {3,7} to fit nicely on a hypothetical smooth surface with the same negative curvature everywhere.

Such surfaces do exist, although there is a catch. Fig. 4.9 shows one of these surfaces, called the *pseudosphere*. It is like the sphere in that it has the same curvature everywhere, but its curvature has the opposite sign, hence the name. This surface would be colored a uniform blue if printed in the same style as in fig. 4.4.

The surface is tiled with triangles (although in a new pattern, the *(7,3,2) triangle tiling*, more on this

Nobody knows whether it is actually possible to keep extending a {3,7} hinged surface outward forever, without having the triangles crash into one another. As the surface grows outward, it acquires more and more ripples. A larger and larger number of triangles need to fit into a growing amount of space, but the number of triangles grows far faster than the amount of space to fit them into. The surface cannot be extended outward indefinitely if the triangles have thickness: each one takes up some fixed amount of space, and eventually you run out of space. But if the triangles have zero thickness, right now nobody knows.

soon), and again I've shaded in blue the corresponding triangle in fig. 4.8. The blue triangle takes up a sizable portion of the pseudosphere (in fact, for reasons I won't go into, precisely one-fourteenth of its area). The catch is that although it seems like our hinged surfaces should go on forever outward, this is all of the pseudosphere that we can ever get. If you try to make more of the surface, extending from its circular boundary at the bottom, then it is impossible to keep the curvature the same on the added surface.

The German mathematician David Hilbert first proved that this is the way it has to be for any smooth surface of constant negative curvature in three-dimensional space: There is no way to extend such a surface indefinitely. Any such surface must have a boundary that we cannot continue past.

Nevertheless, there is a kind of geometry that has constant negative curvature and a way to see it. However, we won't be able to see it without some distortion. The situation is very similar to the problem of being a two-dimensional person trying to understand a three-dimensional object using projection, as in chapter 3. We will be projecting a geometric object from another space down into our space.

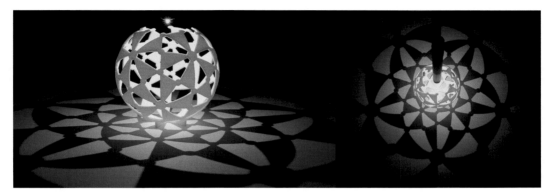

Let's warm up by looking again at stereographic projection from the sphere to the plane, and a new kind of tiling we just saw in fig. 4.9: triangle tilings. Fig. 4.10 shows the *(5,3,2) triangle tiling* of the sphere (together with its stereographic projection to the plane), and fig. 4.11 shows the closely related *(6,3,2) triangle tiling* of the plane.

Unlike Schläfli symbols (which talk about tilings in all dimensions), triangle tilings only talk about two-dimensional tilings made out of triangles. The numbers tell us how many triangles meet at the vertices—just double the three numbers. Or paying attention to the two colors of triangle in the pattern, the numbers say how many triangles of a single color meet at each vertex. Or another way to think about it is that the numbers count how many planes of mirror symmetry pass through the vertices: the triangles in fig. 4.10 are precisely the panels of the *532 symmetry type, as in fig. 1.29, ignoring the little "buttons" that hold the 3D print together. So this pattern on the sphere has symmetry type *532, if we ignore the fact that half of the triangles are solid and half are empty and treat them all the same. If we do pay attention to the distinction between solid and empty triangles,

Fig. 4.10. Stereographic projection of the (5,3,2) triangle tiling, from the sphere to the plane.

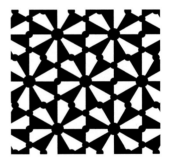

Fig. 4.11. The (6,3,2) triangle tiling lives on the euclidean plane.

The (5,3,2) triangle tiling lives on the sphere, and the (7,3,2) triangle tiling lives on the hyperbolic plane. In between these two, the (6,3,2) triangle tiling lives on the euclidean plane. Triangles in the euclidean plane have angles that add up to 180 degrees. We can check this here: the angles are 180/6, 180/3, and 180/2, and because $1/6 + 1/3 + 1/2 = 1$, the sum of the angles is 180 degrees. For spherical triangles, the angles add up to more than 180 degrees: for the (5,3,2) triangle tiling, we get $1/5 + 1/3 + 1/2 = 31/30$. For hyperbolic triangles, the angles add up to less than 180 degrees: for the (7,3,2) triangle tiling, we get $1/7 + 1/3 + 1/2 = 41/42$.

then the symmetry type becomes 532 (in both cases ignoring the hole at the top where the light goes).

We haven't talked about symmetry on the plane, only on the sphere, but many things work out in much the same way. Symmetry types can again be identified by points with rotational or kaleidoscopic symmetry. The (6,3,2) triangle tiling has symmetry type *632 if we don't care about the colors of the triangles, and 632 if we do.

Just like we saw in chapter 3, the stereographically projected shadow of the (5,3,2) triangle tiling is a distorted version of the true object. Many of the symmetries are lost, and distances between things that we see in the shadow are not the same as the true distances between them. But we can still understand the true object as long as we are aware of how things are distorted.

The true space with constant negative curvature is called the *hyperbolic plane*. (From now on, I'll refer to the ordinary plane as the *euclidean plane* to avoid confusion.) Fig. 4.12 shows three different shadows, or *models*, of the hyperbolic plane, patterned with the (7,3,2) triangle tiling. Just as when we projected four-dimensional things to our three-dimensional

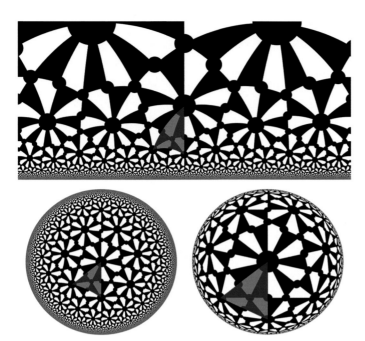

space so we could see them, there is a true object (the hyperbolic plane) that has the correct symmetries and lengths, but we can only see distorted shadows of it. All of the triangles are really the same size and shape in the hyperbolic plane, and it is only the distortion of the projection that makes them look like they're getting smaller toward the boundary. This is analogous to the situation with the (5,3,2) triangle tiling in fig. 4.10: on the sphere all of the triangles are the same size and shape, but in the projection they get bigger the farther we go out, corresponding to getting closer to the north pole of the sphere. An ant walking around on the sphere could walk toward and even right over the north pole and wouldn't notice anything unusual. But an observer that could only see the shadow on the plane would see something strange: the ant would appear to get bigger and bigger as it moved farther out on the plane (i.e., toward the north pole). It has to get bigger and bigger, and the lengths of its steps have to get bigger and bigger, because the ant takes the same number of steps to walk over each of the triangles it sees on the sphere. So it has to get infinitely far out on the plane in only finitely many steps.

For the hyperbolic plane, again we can only see

Fig. 4.12. Different "shadows," or "models," of the hyperbolic plane. *Top*, the upper half plane model; *bottom left*, the Poincaré disk model; *bottom right*, the Klein model. One of the triangles from the {3,7} tiling is highlighted in blue in each model.

the shadows, we don't get to see the true object. As an ant walking around in the hyperbolic plane gets toward the boundary of the shadow, it would appear to get smaller, so that it would take the same number of steps to walk across each of the triangles. In fact, it would never get to the boundary, because there are infinitely many triangles between it and the boundary—the boundary is infinitely far away.

I can't show you the true hyperbolic plane itself. This time, the problem is not that it lives in a higher dimensional space; instead, the problem is that it lives in a space with a different sense of distance. Getting into this properly is beyond the scope of this book, so I'll just pique your interest: one way to think about the true hyperbolic plane is as a sphere of radius $\sqrt{-1}$, living in a relativistic space-time continuum.

In any case, we can understand what's going on by looking at the shadows—the different models shown in fig. 4.12. The *Poincaré disk model* (bottom left) and the *Klein model* (bottom right) both fit inside a disk. As you might guess from the name, the *upper half plane model* occupies the whole upper half of the plane, and so I had to cut it off at the top and sides in fig. 4.12. It isn't so easy to fit half of an infinite plane into a book, although the Poincaré disk and Klein models are really the same thing as the upper half plane so I guess I already did. I've colored one of the triangles from the {3,7} Schläfli symbol tiling in blue in fig. 4.12. You can check that seven such triangles fit around each of the three vertices of the blue triangle in each of the three models. The Poincaré disk model is named after Henri Poincaré, and the Klein model after Felix Klein, but all three of these models were first proposed by Eugenio Beltrami. People don't always get the credit they deserve.

Fig. 4.13 shows how the three different models of the hyperbolic plane are related to one another, using yet another model: the hemisphere model. Let's think of this as the southern hemisphere of a sphere. If we put a point light source at the north pole of this sphere, then the shadow you get is the Poincaré disk

model. For the Klein model, move the light source very far away upward (really, the light rays should be coming down parallel to each other). For the upper half plane model, the light source goes on the equator of the sphere, as shown in fig. 4.14.

Just like stereographic projection from the sphere to the plane, both the Poincaré disk and upper half plane models also preserve angles. That is, the angles we see at the corners of the triangles are the true angles in the hyperbolic plane. They are not distorted, although lengths obviously are. The angles in the Klein model are true at the very center but get more and more distorted toward the boundary of the disk. Don't worry, the Klein model beats the other two in a different way.

The sides of the triangles of the (7,3,2) tiling are straight lines in the hyperbolic plane. Just as in the

Fig. 4.13. Relations between the three different models of the hyperbolic plane via a fourth model: the hemisphere model.

Fig. 4.14. Putting a point light source at the equator of the hemisphere model casts the upper half plane model as a shadow.

We need to know what "straight line" means on the sphere or on the hyperbolic plane. We need lines to draw triangles, for example. On the euclidean plane, the shortest path between two points is a straight line. The thing to do for the sphere and the hyperbolic plane is to flip this around and say that a "straight line" is the shortest path between two points: pull a string tight and see what you get. On the sphere, you get a part of a *great circle*—the biggest kind of circle you can draw on a sphere. For example, the equator is a great circle. We can describe what the straight lines are on the hyperbolic plane in different ways for the different models (see fig. 4.12). The Klein model shines here: The straight lines are just ordinary straight lines on the ordinary euclidean plane that we draw the Klein model on. For the Poincaré disk model, the straight lines are either ordinary straight lines through the center of the disk, or arcs of circles that meet the boundary at right angles. For the upper half plane model, we get vertical straight lines, and semicircles meeting the boundary at right angles.

euclidean plane, we can reflect across a straight line. For the (7,3,2) triangle tiling, reflecting across one of the edges switches the black and white triangles, but otherwise doesn't change the tiling. This is easy to see if you reflect across one of the lines that looks straight in any of the three models, but in fact it's true for every single line. Every tiny semicircle in the upper half plane model has exactly as much stuff on one side of it as on the other. It has to since you can reflect across it to switch one half with the other. As with the (5,3,2) and (6,3,2) tilings, the (7,3,2) tiling has a symmetry type, but this time, it is a symmetry of the hyperbolic plane rather than the sphere or the euclidean plane. Just as before, the symmetry type is *732 if we don't care about the colors of the triangles, and 732 if we do.

The box above describes what a "straight line" is on the hyperbolic plane. I haven't talked about it here, but it is also possible to say what circles are. With circles and lines we can start doing the kind of ruler and compass geometry Euclid wrote about. Euclid started with four very basic postulates, and a fifth not-so-basic postulate. Very briefly, Euclid's first four postulates say things like "you can draw a line between any two points" and "you can draw a circle centered on a point that goes through another point." The fifth postulate is considerably more complicated but boils down to the following: "Given a line and a point not on the line, there is precisely one other line that intersects the point and does not intersect the line."

Euclid realized that his fifth postulate seemed very complicated in comparison to the other four and proved as much as he could about geometry without using it. Many people through history attempted to prove the fifth from the other four, hoping that they could clean up this apparent ugliness at the core of euclidean geometry. This search was futile. In 1823, Janos Bolyai and Nicolai Lobachevsky independently discovered *non-euclidean geometries*, in which Euclid's first four postulates hold, but the fifth does not. The sphere and the hyperbolic plane are two of these. On the sphere, if you have a line (a great circle) and a point not on it, then any great circle going through that point has to intersect the first great circle. On the hyperbolic plane, it's the other way around: if you start with a line and a point not on it, you can find many lines that go through the point and don't intersect the first line. See if you can find some examples of this in the (7,3,2) triangle tiling.

We can now say where *all* of the length-two Schläfli symbol tilings live. Fig. 4.15 is a map of the terrain. There are five Schläfli symbols that correspond to the regular polyhedra and so really live on the sphere: {3,3}, {4,3}, {3,4}, {5,3}, and {3,5}. Three more are tilings of the plane: {3,6}, {4,4}, and {6,3}. We have only looked carefully at {3,7}, but it turns out that there's enough space on the hyperbolic plane to fit in *all* of the other tilings.

In fact, the hyperbolic plane can fit in even more Schläfli symbol tilings than you might think. Fig. 4.16 shows the Poincaré disk model tiled with the *(∞,3,2) triangle tiling*. That is, at the corners of each triangle, either two, three, or infinitely many triangles of a single color meet. Putting six of these triangles together gives a triangle (the highlighted blue triangle) from the {3,∞} Schläfli symbol tiling. To fit infinitely

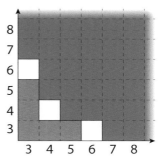

Fig. 4.15. All of the length-two Schläfli symbols are tilings on either the sphere (*red squares*), the euclidean plane (*white squares*), or the hyperbolic plane (*blue squares*; extends upward and to the right forever).

Fig. 4.16. The (∞,3,2) triangle tiling in the Poincaré disk model. The highlighted blue triangle is a tile of the {3,∞} Schläfli symbol tiling.

many triangles around a vertex, it turns out that the vertex has to be infinitely far away, which is why the corners of the blue triangle go out to the boundary of the Poincaré disk model. This is very different from the euclidean plane. In the euclidean plane, you can make a line whose ends are infinitely far away, but you cannot make a triangle whose vertices are infinitely far away. (What happens if you try?)

You might be getting the sense that there is just in general more going on in the hyperbolic plane than for the sphere or the euclidean plane. And you would be right: There are more lines going through a point that don't intersect another line. There are (infinitely more) Schläfli symbol tilings, and there are many, many more ways to be symmetrical.

Three-Dimensional Tilings

In two dimensions, we looked at tilings given by Schläfli symbols on the sphere, the euclidean plane and the hyperbolic plane. We already saw some length-three Schläfli symbols: the regular polychora—the hypercube and its friends. The beach ball versions of the polychora (as in chapter 3) are all tilings of the three-sphere. What about tilings of three-dimension-

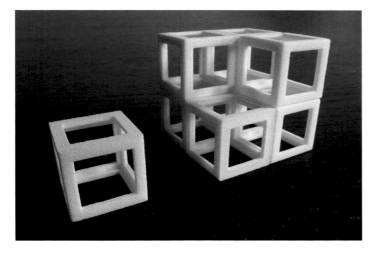

Fig. 4.17. Cubes stack together to tile space.

al euclidean space or three-dimensional hyperbolic space? (There is a three-dimensional version of the hyperbolic plane. We'll get there in a minute.)

There is only one length-three Schläfli symbol that tiles three-dimensional euclidean space: {4,3,4}. This is a very familiar tiling, cubes stacked with four around each edge fill up space. See fig. 4.17. The angle between two faces of a cube is 90 degrees, so four cubes fit perfectly around an edge. There aren't any other ways to tile euclidean space with regular polyhedra, because none of the other regular polyhedra have an angle between their faces that evenly divides 360 degrees (see fig. 3.16).

Nonregular Tilings of Three-Dimensional Euclidean Space

If we don't require all of the tiles to be the same regular polyhedron, then there is another thing we can do: although octahedra and tetrahedra cannot tile euclidean space on their own, together they can. See fig. 4.18. The pattern is a little confusing. One way to understand what's going on is shown on the right of fig. 4.18. Two tetrahedra fit on either side of an octahedron to make a skewed cube. Then this skewed

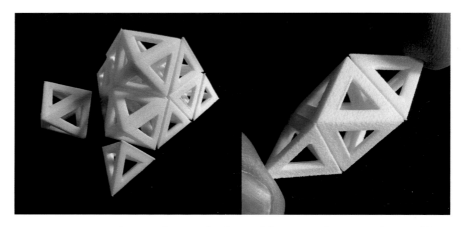

Fig. 4.18. Regular octahedra and tetrahedra can together tile space together, in a ratio of 1 to 2.

cube can be stacked just like an ordinary cube to fill space.

Alternatively, we could break the rules by also allowing archimedean polyhedra (as in fig. 2.9). There is one archimedean polyhedron that tiles space on its own, the truncated octahedron (see fig. 4.19).

If we allow multiple kinds of archimedean polyhedra, then there are many more possibilities. One of the prettiest uses cubes, truncated octahedra, and truncated cuboctahedra. Can you figure out how they fit together?

If we also allow non-archimedean polyhedra then there are many more ways to fill space. My absolute favorite polyhedron that tiles space is the *rhombic dodecahedron* (see fig. 4.20). It is a dodecahedron, meaning that it has 12 faces, but unlike the regular dodecahedron, the faces are rhombuses rather than pentagons. Bees also like the rhombic dodecahedron. Their honeycombs are made from cells, each of which is a hexagonal prism (as seen in fig. 2.8), capped off at one end by half of a rhombic dodecahedron. The molecular structure of diamond is also closely related to the tiling of space by rhombic dodecahedra. Each carbon atom sits at a vertex of the tiling, and the four

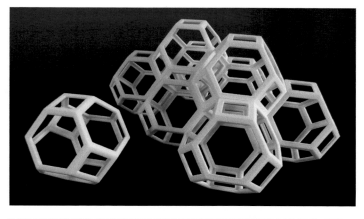

Fig. 4.19. Truncated octahedra tile together to tile space.

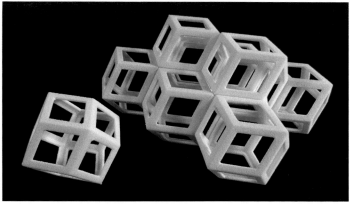

Fig. 4.20. Rhombic dodecahedra tile space.

connections from each carbon atom to its neighbors run along the edges of the tiling. All of the vertices where four faces of a rhombic dodecahedron meet and half of the vertices where three faces meet have a carbon atom.

Many other polyhedra that tile space have been discovered. How many? Well, one easy way to make any number of new space-tiling (also called *space-filling*) polyhedra is to just take a cube and make some kind of bump on one side and a corresponding dent on the other. This is too easy though, so to make it an interesting problem we could require that the polyhedra are convex, without any dents. This rules out bump and dent trickery. Nobody knows what the complete list of these looks like. We don't even know how many faces a convex polyhedron that tiles space can have. So far examples with up to 38 faces have been found. But maybe there are even more complicated examples out there to find. Here's an easier question to ponder:

can you find a nonregular tetrahedron that tiles space? Aristotle thought that the regular tetrahedron does, but it doesn't, which is obvious if you have a few of them to play with (see fig. 3.16). But what if we look at tetrahedra that have edges of different lengths?

Tilings of Three-Dimensional Hyperbolic Space

In fig. 4.15, we saw that of the two-dimensional regular tilings there are five spherical, three euclidean, and then everything else is hyperbolic. A similar thing happens in three dimensions. There are six spherical tilings (the regular polychora, as we saw in table 3.1), only one euclidean tiling, and everything else is hyperbolic. That is, *all* of the length-three Schläfli symbols other than $\{3,3,3\}$, $\{4,3,3\}$, $\{3,3,4\}$, $\{3,4,3\}$, $\{5,3,3\}$, $\{3,3,5\}$, and $\{4,3,4\}$ live in three-dimensional hyperbolic space.

Just as for the hyperbolic plane, we cannot see three-dimensional hyperbolic space directly, so we need to look at projections of it into our space. There are versions of the Poincaré disk model, the Klein model and the upper half plane model for hyperbolic space. Let's just look at the three-dimensional version of the Poincaré disk model, namely, the *Poincaré ball model*. The Poincaré disk model lives in a disk inside of a circle, while the ball model lives in a ball inside of a sphere.

Let's look at an example. Fig. 4.21 shows the $\{5,3,4\}$ Schläfli symbol tiling, a *hyperbolic honeycomb*. This is a complicated object, and it's difficult to see what's going on from the photographs. (If you don't have a 3D print of this, I encourage you to check out the virtual 3D model online at 3dprintmath.com.) This is a ball, tiled with dodecahedra. Four dodecahedra meet around each edge. Just as with the tiling of the two-dimensional Poincaré disk model, the dodecahedra get smaller as we get nearer to the surface of the ball, which just like the circular boundary of the Poincaré disk, is infinitely far away. The edges of the dodecahedra are straight lines in hyperbolic space, that is, shortest paths between points, and just as for

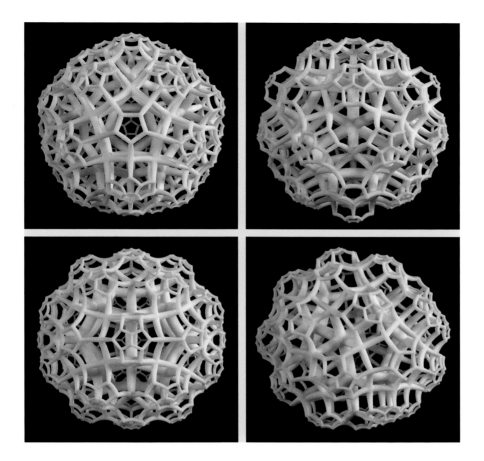

the Poincaré disk model, these are arcs of circles that hit the boundary at right angles.

I worked on this visualization (and others) with Roice Nelson, a computer programmer from Austin, Texas. We couldn't tile that far out toward the boundary because the edges of the dodecahedra would get too thin to print. Despite how complicated this looks, there is only one dodecahedron in the middle, its 12 neighbors and some of the neighbors of those neighbors. But the full tiling extends outward infinitely, getting smaller and smaller as it fills up the ball.

I'll show you just a couple more hyperbolic honeycombs. First, fig. 4.22 shows the return of the icosahedron in the {3,5,3} honeycomb. We have seen all of the other regular polyhedra appear as cells of polychora, but the icosahedron can only be a cell of a regular tiling in hyperbolic space. Here, three icosahedra fit around each edge. This model has even fewer cells

Fig. 4.21. The {5,3,4} hyperbolic honeycomb, drawn in the Poincaré ball model.

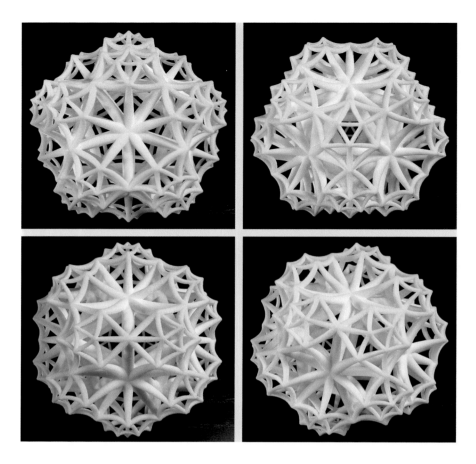

Fig. 4.22. The {3,5,3} hyperbolic honeycomb, drawn in the Poincaré ball model.

Fig. 4.23. The {3,3,6} (*opposite, above*) and {6,3,3} (*opposite, below*) dual hyperbolic honeycombs, drawn in the Poincaré ball model.

than the {5,3,4} honeycomb: there is only one icosahedron in the middle and its 20 neighbors, but again the full tiling goes outward forever, getting smaller and smaller as it fills up the ball.

Finally, fig. 4.23 hints at how much more weirdness there is out there in hyperbolic space. At the top is the {3,3,6} honeycomb, made out of tetrahedra (because the first two numbers are {3,3}), with six around each edge (the last number). So far so good. But remember from chapter 3 that you can get the dual of a polytope by reversing its Schläfli symbol. If we do that here, we get {6,3,3}, as shown at the bottom of fig. 4.23. What is this made out of? The cells are supposed to be given by the first two numbers of the Schläfli symbol, so the cells should have Schläfli symbol {6,3}. But that isn't a polyhedron; it's the tiling of the euclidean plane by hexagons with three around each vertex. Well, it turns out that this just works anyway: the hexagonal

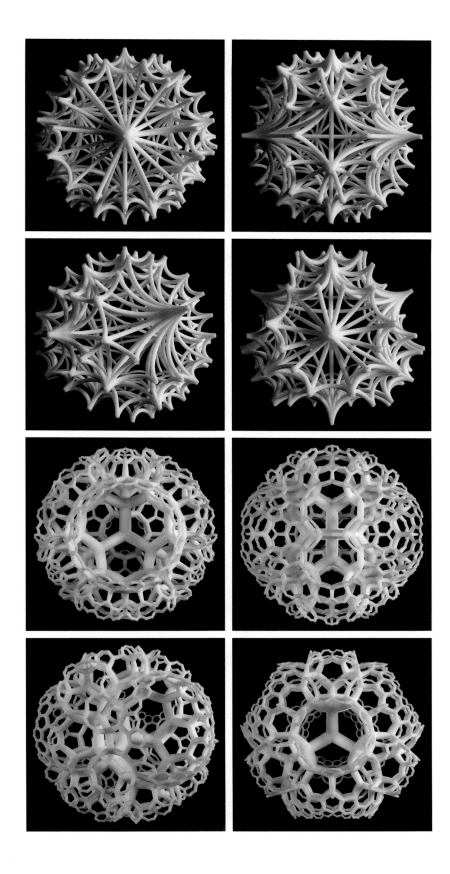

tiling gets pasted onto a sphere just like a polyhedron, except that this sphere touches the boundary of the Poincaré ball model, infinitely far away. It has to in order to fit the whole infinite {6,3} tiling onto it. The vertices of {3,3,6} should be at the centers of the cells of {6,3,3}, and they are. As with the triangles of Schläfli symbol tiling {3,∞} in fig. 4.16, all of the vertices of these tetrahedra are also infinitely far away, on the boundary of the Poincaré ball model.

There is much more to this story of fitting tilings into three-dimensional hyperbolic space and how to see what they look like. For more details, see my paper with Roice Nelson, "Visualizing Hyperbolic Honeycombs" (details in appendix A, fig. 4.21).

5 Knots

Knots arise naturally in power and headphone cables, and artificially when we use them to tie up trash bags, clotheslines, sails, and other things. They also show up in biology: the DNA molecules in our cells are very long, flexible strands, and these can get knotted. There are also some surprising connections between the mathematical theory of knots and such esoteric topics as quantum field theory and quantum computation.

Is it always possible to untie a knot? As long as the knot isn't tied so tightly that you can't move the rope then, yes, it is always possible to untie. You just have to find a free end and pull it back through everything else. Given enough time, the knot will eventually come undone. If we had an idealized piece of string, a perfectly flexible curve in space with no thickness and no friction, then this process would always undo any knot. Mathematicians like to study these kinds of idealized objects, because the real world is often far too complicated and messy to be able to prove theorems about. In the case of knots, however, it looks like there's nothing to prove: an idealized knot can always be untied. The trick to make it interesting is to tie a knot and then fuse the ends of the string together so

Fig. 5.1. The unknot (*left*), the trefoil (middle), and the figure-eight knot (*right*).

that it forms a closed loop. Fig. 5.1 shows a few possibilities for what you might get after doing this.

For the first loop in fig. 5.1, I didn't actually tie a knot at all, giving the *unknot*. For the second, I made an overhand knot, or *trefoil knot*. The third is the *figure-eight knot*. These are particularly simple pictures of these knots. There are lots of other pictures of the same knots. For example, I could put an extra twist in the trefoil, to get fig. 5.2. In chapter 1, we talked about taking different photographs of objects by moving the camera around them. There the objects we were looking at were rigid. But here the objects are flexible. We get different photographs not only by moving the camera around but also by deforming the knot. So, there is a truly enormous variety of different ways to take a photograph of any particular knot.

The way I've named these three knots suggests that I think that they are different knots. But is even that so obvious? Perhaps we could somehow "untie" the trefoil knot so that it becomes the unknot, and then the two pictures would really be two pictures of the same knot. We could play around with it—pass strands through and around loops, add and remove twists—perhaps if we play for long enough it will

Fig. 5.2. Another photograph of the trefoil knot.

Fig. 5.3. Is this also the trefoil knot?

all come undone and we will just get the unknot, a circle.

So there is an interesting problem: how can we tell if two photographs of knots are photographs of the same knot, or of different knots? Sometimes it is easy. Fig. 5.2 obviously shows the same knot as the trefoil in fig. 5.1. Sometimes it isn't so easy. Is the knot pictured in fig. 5.3 also the same?

There is a useful analogy here with fractions. The fractions 1/2 and 2/4 are different "photographs" of the same number. We can easily tell whether two fractions are the same number, because we have a standard, best "photograph": divide out any common factors between the numerator and the denominator. If we want to find out whether 39/63 is the same as 52/84, we just put both of them in this best form and discover that they are both the same as 13/21. Unfortunately, with knots, there doesn't seem to be a best photograph we can go to. We have moves to change one picture of a knot into another picture of the same knot (e.g., by adding or removing a twist), just as dividing the numerator and denominator of a fraction by the same number is a move that doesn't change the number. But we don't have a best destination to go to or an obvious way to get there.

One of the motivating questions in the mathematical study of knots is, How can we tell if two pictures of knots are really the same knot, or are really different knots? This is really two different questions:

1. How can we tell if two pictures of knots are really the same knot?
2. How can we tell if two pictures of knots are really different knots?

The first question is easier to answer, in a way. If we have two pictures of the same knot, then all we have to do is show how to move the knot in one picture around to get to the other. The second question seems much harder at first. How could you possibly show for sure that there is no way to turn one knot picture into another? Even if you tried for weeks and didn't get anywhere, maybe if you just tried for five minutes longer you might succeed? Amazingly, it is possible to prove for sure that there is no way to turn the trefoil knot into the unknot without cutting the string and reattaching it. In fact, there are many ways to prove it. I won't get into the details here. I'll just recommend that if you're interested you should look up *tricolorability* for a first way to prove it and mention that it really isn't that hard to follow the argument. I can unreservedly recommend *The Knot Book* by Colin Adams for much more on this.

Geometry versus Topology

The question of whether two pictures of knots are the same is not a geometric problem in the usual way we think about geometry. The specific shapes of the knots don't matter to the question. If we nudge one of the knot pictures a bit, it doesn't make any difference to the question, even though it changes the geo-

metric shape of the knot. Rather, this is a *topological* question. *Topology* is the study of geometric objects when we don't care about lengths or angles. Everything is made from infinitely stretchy and flexible rubber, and all that matters is how things are connected to other things.

There is a standard joke about topologists: that they can't tell the difference between a doughnut and a coffee mug (see fig. 5.4). Both the coffee mug and the doughnut have a hole—the hole through the doughnut and the hole through the handle of the coffee mug. The coffee mug also has a big indentation where the coffee goes, but in the world of topology, everything can be deformed as much as we want, as long as we don't cut or tear anything. We can gradually deform away the coffee-intended indentation to turn the mug into the doughnut.

In my day job, when I'm not writing books about 3D printed mathematical things, I do research in topology. For me, at least, this involves lots of sitting down and drawing pictures of topological things. But any drawing (or, for that matter, 3D print) is a physical object in the real world, so it has some actual geometry, even if the precise geometry doesn't matter

Fig. 5.4. A very flexible coffee mug can be deformed into the shape of a doughnut. Or read in the other direction, a very flexible doughnut can be deformed into the shape of a coffee mug. (Or so I'm told. I can't tell the difference between any of these things.) This sequence of shapes was designed/generated by Keenan Crane.

to a topological question. And we have to choose that geometry somehow.

There are approximately three main strategies for choosing the geometry of a topological object, which I call the *by-hand* method, the *parametric/implicit* method, and the *iterative* method.

The by-hand method is what you use when you draw a picture of a knot by hand or make it out of string. For a 3D printed design, this would be using the user interface of a computer program to build the geometry or (as I did in figs. 5.1 through 5.3) arranging a flexible "pop-beads" model by hand. With the parametric/implicit method, the geometric shape is determined by some mathematical formula or a procedure that directly produces the geometry. With the iterative method, we start with a design produced either by-hand or parametrically, and then use a computer program to gradually alter the design. The idea is to improve how "good" the design is by nudging it, slowly evolving to a better design. Of course, we have to say what "good" means, and a different way to measure "goodness" will give a different final result.

Let's see some examples of these different ways of doing things. First, let's look at one way to parametrize the trefoil knot, based on a surface called the *torus*. This is the same as the surface of a doughnut (see fig. 5.4). Fig. 5.5 shows another way to think about the torus. Here I've made a model of the torus out of squares. The squares form rings 10 squares long in each of the two directions. If we cut the torus apart along the green circle and unroll the tube, we get a cylinder. Then we can cut along the blue line as well and unroll that into a square. Just to be clear, cutting is not a legal thing to do to a topological object. The torus is not the same as a square. But this does give us a good way to think about the torus. We can make one by starting with a square, rolling it into a tube (gluing the blue edges of the square together), and then rolling that tube into the doughnut shape (gluing the green circles at the ends of the tube together to make the doughnut).

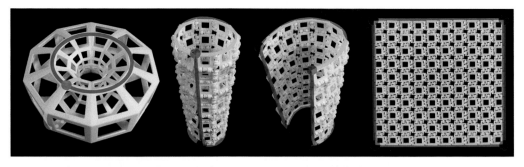

Now let's draw some lines on the square in such a way that they will join up when we roll it up to make the torus. See fig. 5.6. The lines join up if, whenever a line ends somewhere on the right side of the square, another line starts at the corresponding point on the left side of the square. Then when the square is rolled up into the torus, the lines connect together. The same "matching up" also needs to happen on the top and bottom edges of the square. If we draw our lines on the square in the right way, we can make the trefoil knot (see fig. 5.6, *right*).

As a fun exercise, you might want to try this with a square of paper and some string. Tape four pieces of string onto the square of paper as in fig. 5.6 (*left*) to keep them in place. Roll the square up into a tube, and tape the string together where they meet. Then roll the tube into a torus (this has to involve some crumpling or folding), and again tape the string together where they meet. Cut away the paper, and you should get the trefoil. Alternatively, wind the string around a doughnut or bagel, tape the ends together, and then eat away your toroidal baked good of choice.

This process gives us a precise way to make the trefoil knot (in fact, a parametrization), assuming that

Fig. 5.5 The torus (*top, left*) can be cut and unrolled into a cylinder, and cut once again to be unrolled into a square.

Fig. 5.6. The lines on the square (*bottom, left*) turn into the trefoil (*bottom, right*) when the square is rolled up to make the torus (*bottom, middle*).

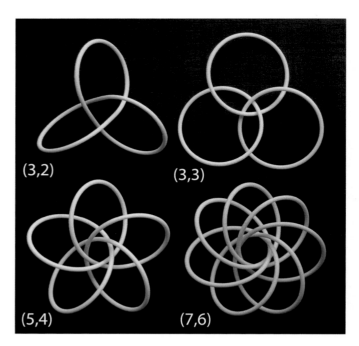

(3,2) (3,3)

(5,4) (7,6)

Fig. 5.7. Three torus knots and a torus link.

we have a precise way to make a torus. The "rolling up a square" idea can be turned into a parametrization of the torus using some trigonometry (details in appendix A).

The trefoil is an example of a *torus knot*. Torus knots are the knots that you can make by drawing on a torus, like we did for the trefoil. We can make different torus knots by going different numbers of times around and through the torus. The trefoil is the *(3,2) torus knot,* because it goes three times through the hole and two times around. In the unrolled version in fig. 5.6, the knot goes three times around horizontally, and twice vertically.

Fig. 5.7 shows what you get from some other pairs of numbers. Something quite different happens for the pair (3,3). Rather than a knot, we get a *link* made up of three loops. Links are just collections of knots arranged together in some way. For the (3,3) torus link, all three of the knots are copies of the unknot. Here's a puzzle for you to ponder: Which pairs of numbers will give a torus knot, and which give a torus link, made up of more than one knot? If we get a link, how many knots is it made out of?

These are very nice, symmetrical ways to position

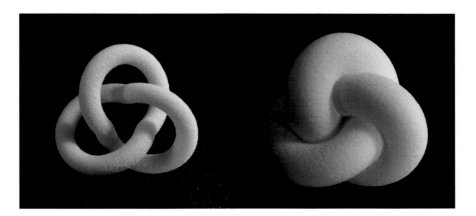

torus knots in space. They could have had no symmetry at all, like the trefoil in fig. 5.2, for example. The trefoil in fig. 5.7 has an obvious threefold rotational symmetry, but it also has a less obvious symmetry. If you turn it over on the table, it looks the same again. In fact, in the language of chapter 1, it has symmetry type 223. (Another puzzle: what about the other torus knots?)

Fig. 5.8. *Right*, a minimal ropelength trefoil knot. *Left*, the same shape of knot made with a rope of half the width.

These torus knots are examples of parametric shapes for knots. Next let's look at some knot shapes made using an iterative method. Fig. 5.8 shows two pictures of the trefoil, generated by Jason Cantarella, Eric Rawdon, Michael Piatek, and Ted Ashton. On the right is a *minimal ropelength* trefoil knot. This is the shape you get if you tie a knot as tightly as possible, using a physical piece of rope with a fixed thickness. This shape was produced by the computer program Ridgerunner, which simulates slowly pulling a knot tight, reducing its length while making sure that it isn't allowed to overlap itself in space. On the left of fig. 5.8 is the same core knot shape, made with a tube with half the thickness. This makes it easier to see what's going on. I should say that, at the time of this writing, there is no proof that these shapes use the

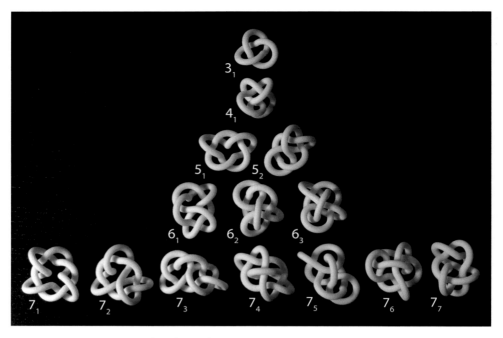

Fig. 5.9. Prime knots with up to seven crossings, with (near) minimal ropelength shapes.

absolute shortest possible length of rope. They are just what the computer simulation finds. Maybe there is some slightly different way to tie it that is very slightly shorter, but nobody knows for sure.

Fig. 5.9 shows some other knots with shapes calculated by pulling them tight. This is the start of the list of all *prime knots*. A prime knot is a knot that cannot be made by combining two smaller knots, just as a prime number is a number that cannot be made by combining two smaller numbers. For numbers, *combine* means multiply. For knots, *combine* means doing something called a *connect sum*. Fig. 5.10 shows the idea. Take two knots arranged so that they almost touch each other. Break the knots apart and then connect them to each other instead, making one knot. Any whole number can be split into prime numbers in only one way. Likewise, any knot can be split into prime knots in only one way. (Can you see why? This is a bit tricky.)

Fig. 5.9 shows the 14 simplest prime knots. "Simplest" here means in terms of the smallest number of crossings of the knot. A *crossing* in a photograph of a knot is a place where one strand crosses over another.

Different photographs of a knot can have different numbers of crossings. For example, the photograph of the trefoil in fig. 5.1 has three crossings, while the photograph of the trefoil in fig. 5.2 has four crossings. But there is always a smallest possible number of crossings in a photograph of each knot—the rows of fig. 5.9 show the knots that can be photographed with three, four, five, six, and seven crossings minimally. Not all of the pictures of knots in the figure actually have the minimal number of crossings. For example, the knot 4_1 has five crossings as photographed in fig. 5.9, but the minimum number of crossings is four. In fact, 4_1 is the figure-eight knot, as seen in fig. 5.1. I'll leave it to you to work out how to move the knot around from the picture in fig. 5.9 to make it look like the figure-eight knot shown on the right of fig. 5.1.

Incidentally, the labeling in fig. 5.9 is the standard notation used in knot theory: it gives the minimum number of crossings as the main number, with a subscript to distinguish between knots with the same minimum number of crossings. But there is no special meaning to the ordering of those knots.

The torus knots are all prime knots, and so they

Fig. 5.10. Two trefoil knots (*left*) and the result of taking their connect sum (*right*).

Before we knew that atoms are made out of protons, neutrons, and electrons, the British mathematical physicist Lord Kelvin theorized that atoms were knots, tied in the ether. (We didn't know that the ether doesn't exist either.) So, people started trying to catalog knots, hoping to match them up with the different chemical elements. Then the physicists figured out that atoms didn't have anything to do with knots and moved on. But, by this time, the mathematicians were hooked and have been classifying knots ever since. This gets difficult to do as you look at more and more complicated knots. If I were to extend the table of knots shown in fig. 5.9, the next row would have 21 knots, then 49, then 165. The fourteenth row (that is, the knots with minimum crossing number 16) would have 1.3 million knots in it, for a total of just over 1.7 million prime knots with minimum crossing number at most 16.

all appear in the list of prime knots: 3_1 in fig. 5.9 is the trefoil, which is the same as the (3,2) torus knot, 5_1 is the same as the (5,2) torus knot, and there's one other torus knot hiding in fig. 5.9 that I'll leave for you to find. This isn't so easy to do, because the minimal ropelength knot shapes often have less symmetry than we might expect. For example, we might expact that the minimal ropelength shape for the knot 5_1, would have fivefold rotational symmetry (in fact, symmetry type 225. But it turns out that it can be tied more tightly by breaking that symmetry.

This is a difference that comes up again and again between the parametric method and the iterative method: a good parametrization often knows something about the topological thing—that it is supposed to have some symmetry, for example. Often the parametric method gives a more beautiful result or tells you more about what's going on. On the other hand, iterative methods always work. We always get some sort of shape out at the end. There might not be a nice simple way to describe a knot (i.e., a parametrization). Once we start looking at knots that aren't torus knots, this starts getting much harder to do.

Fig. 5.11. Two views of a parametrized model of the figure-eight knot.

The figure-eight knot is the simplest knot that isn't a torus knot. Already it is remarkably difficult to come up with a nice parametrization. Fig. 5.11 shows a parametrization of the knot that is as symmetrical as possible, using some trigonometric functions (more details in appendix A).

What is its symmetry type, as in chapter 1? This is tricky. There is an obvious twofold rotation that you can see in the right photo in fig. 5.11, but there are no other obvious rotations or reflections to make, even though the top two "lobes" look similar to the bottom two. Well, I'll leave you to think about it.

There are also tables of all of the possible links. I won't get into these here, but I do want to show one more link, the famous *borromean rings* (see fig. 5.12). The borromean rings are named after the coat of arms of the Borromeo family crest from the fifteenth century, although the link itself is much older: versions of the rings have appeared in many different contexts throughout history. The design has often been used to express the idea of strength through unity, for the following reason: There are three loops (unknots) that are linked together. However, if you cut any one of

Fig. 5.12. Photograph of two borromean rings. *Left*, similar to the Borromeo crest. *Right*, an even more symmetric configuration. I had to make a stand for this print; otherwise, the three loops wouldn't stay in the symmetric position.

the loops off, the other two fall apart from each other. Only with all three together are the rings united.

This is different from the (3,3) torus link in fig. 5.7. If we cut one of its loops, the other two remain linked together. Another puzzle for you to ponder: is there a link like the borromean rings, in that if any one ring is cut, the others fall apart, but with four rings? What about with five or more rings?

6 Surfaces

If you zoom in on a little part of a knot, it looks like a one-dimensional line. Something that looks like a two-dimensional plane when you zoom in is a *surface*. The sphere and the torus are both surfaces that we have looked at before, as is the plane itself. We also saw various patches of surface; for example, the patches of the circular paraboloid, parabolic cylinder, and hyperbolic paraboloid shown in fig. 4.3. But if we think about these patches topologically, forgetting about the differences in geometry, then all of these patches are really the same: each of them can be squashed down to make a pancake shape—a *disk*. Unlike the sphere and the torus, the disk is an example of a surface that has *boundary*—the loop around the outside.

What other kinds of surface are there? We can cut out a hole from a disk to form a ring, or *annulus*, as on the left in fig. 6.1. An annulus is topologically the same as the cardboard tube in the middle of a roll of toilet paper. You can make the shape from a strip of paper by taping the ends together. If instead you give the strip a half twist before taping, you get the *Möbius strip*, named after August Ferdinand Möbius, second

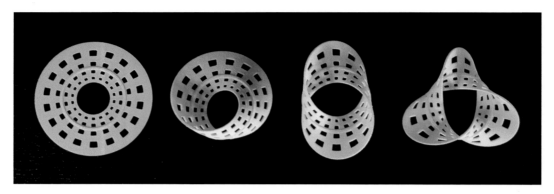

Fig. 6.1. Surfaces made by adding twists to a strip. Each print has a raised circle showing the middle of the strip.

from the left in fig. 6.1. The Möbius strip is famous for having only one side and only one boundary loop.

Or we could give the strip a full twist, or three half twists, as on the right of fig. 6.1. We could keep adding more and more twists if we wanted to.

Intrinsic versus Extrinsic

Imagine that you are an extremely shortsighted ant crawling around on a surface. You can draw on the surface, walk around leaving a trail of breadcrumbs, and so on, but because you are so shortsighted, you can't look out to see how the surface is positioned in space. You would be able to tell the difference between the annulus and the Möbius strip, because you could count that the annulus has two boundary loops while the Möbius strip has only one. You could also tell the difference by painting just one of the two sides of the annulus. This won't work on the Möbius strip, because, of course, it only has one side.

But as a shortsighted ant, you would have no way to tell the difference between the annulus and the full twist strip. While you weren't looking, someone could cut the full twist strip, untwist it, and glue it back together, and you would be none the wiser. The same points of the surface that get cut apart get glued back together, and so any pattern of breadcrumbs you left behind would look exactly the same after cutting and regluing. The shortsighted ant sees the surface *intrinsically*, properties that depend only on the surface itself and not how it is positioned in space. Looking at the surface in space, we see the surface *extrinsically*. From the extrinsic point of view, the annulus and the full strip twist *are* different, and there's no way to deform one into the other, no matter how much you squish it around: The two boundary loops of the annulus are unlinked circles, while the boundary loops of the full twist strip are two linked loops. If you had a way to deform the full twist strip and turn it into the annulus, then you would also have a way to deform the linked boundary loops into the unlinked boundary loops. But that's impossible. (Although it seems obvious that you can't unlink two linked circles, if we are being really careful this can be proved using knot theory.)

From the shortsighted ant's (i.e., intrinsic) point of view, knots are very boring. As far as an ant is concerned, every knot is just a loop: the same as a circle. Things only get interesting when we look at knots extrinsically. On the other hand, surfaces are already interesting from the intrinsic point of view and are a bit too interesting (read: hard) from the extrinsic point of view.

For example, take a look back at fig. 5.11. The print shows a tube representing the figure-eight knot, but we could also think of the tube itself as a surface—in fact a torus. If we think of surfaces extrinsically, then we have to worry about knotted tori. So understanding how a torus can be arranged in three-dimensional space is at least as hard as understanding knots, not to mention all of the other kinds of surfaces.

So, let's stick to the intrinsic point of view and put on our shortsighted ant glasses. For us then, the full twist strip is the same as the annulus. In the same way, the strip with three half twists is the Möbius strip again. The same reasoning shows that any even num-

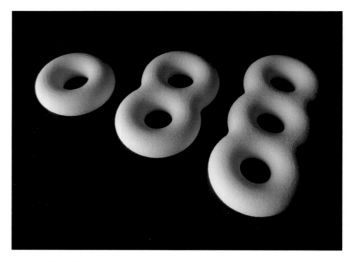

Fig. 6.2. Surfaces of genus one, two, and three.

ber of half twists gives us the annulus, and any odd number of half twists gives the Möbius strip.

We don't get anything new by adding twists to a strip. However, we can get something new by adding *handles* to a surface. Fig. 6.2 shows a torus, what you get by adding a handle to the torus, and what you get by adding a handle again. The number of handles a surface has is called the *genus* of the surface. A torus has genus one and a sphere has genus zero. Surfaces with more handles don't get special names, so the other two surfaces in fig. 6.2 are just called the genus two surface and the genus three surface.

Unlike the Möbius strip, all of the surfaces in fig. 6.2 are two-sided. There's an inside (filled with plastic) and an outside (the rest of the universe). Another famous, one-sided surface is the *Klein bottle*, first described by Felix Klein, and displayed in the form of a bottle opener in fig. 6.3.

The Klein bottle looks something like a torus: a tube is curved around to join up with itself. But instead of connecting up like a torus, the tube crashes through itself to connect up from the other side. This crashing through itself seems somewhat inelegant.

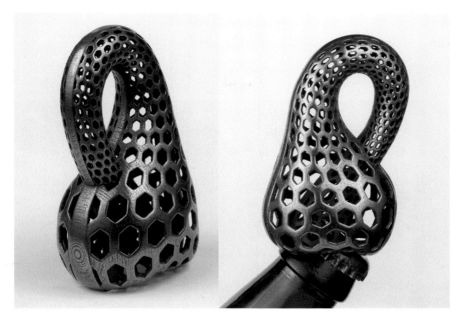

Fig. 6.3. *Klein Bottle Opener*, by Bathsheba Grossman.

Worse than that, is it even a surface? Surfaces are supposed to look like the plane when you zoom in close but is that true if we zoom in where its crashing? In fact, there is no way to put the Klein bottle into three-dimensional space without it crashing through itself: It's a surface without any boundary loops, so it should be dividing space into two parts, except that it can't because it's only got one side. The way out of this conundrum is that it has to crash through itself.

However, remember that we are wearing our shortsighted ant intrinsic glasses. From the perspective of the shortsighted ant, it only sees the surface nearby where it is, not how it is positioned in space. So it doesn't "see" the surface crashing at all. It happily wanders along the tube, not noticing that some other part of the surface, very far away from it "as the ant crawls" is in the same place "as the crow flies." The upshot is that from the intrinsic point of view, there is no crashing (and so, yes, the Klein bottle is a perfectly good surface). It's just an artifact of trying to fit the surface into three-dimensional space so that our poor, three-dimensional brains can see it. It's the same kind of situation as in fig. 3.4. The true object doesn't have

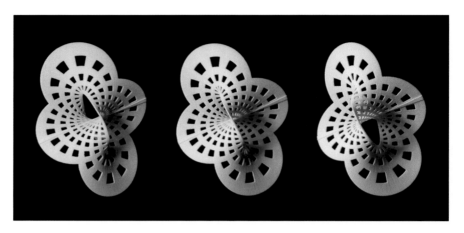

Fig. 6.4. Three views of a mysterious surface.

parts crossing through one another, it's just that our picture makes it look like it does. For the cube in fig. 3.4, we could fix our picture by changing how we projected to two dimensions. For the Klein bottle, there is no way to fix the crashing in three dimensions. However, if you're willing to go up to four dimensions, then everything works out. We can just push the crossing parts away from one another a little bit in the fourth dimension to remove the crashing.

A Mystery Surface

We have seen some different kinds of surfaces, with different properties, either one sided or two sided and with different numbers of boundary loops and with different numbers of handles. Just as with knots, there is a problem that comes up, in how to tell whether two pictures of surfaces are actually the same surface, shown in two different ways. As an example, what is the surface in fig. 6.4? Is it the same as something we have seen before?

First, there are two different boundary loops: there is a circular boundary loop in the middle, and a strange loop around the outside with right-angle turns. With a little more work, you should be able to see that this surface is one sided: find a path to trace your finger along that gets you to the other side from where you started, without going over a boundary loop. So this isn't the annulus, because the annulus is two sided, and it isn't the Möbius strip, because the Möbius strip has only one boundary loop. But this mystery surface is very closely related to the Möbius strip. On the next page, I'll show you exactly how they are related, so if you want to try to figure it out for yourself, stop here.

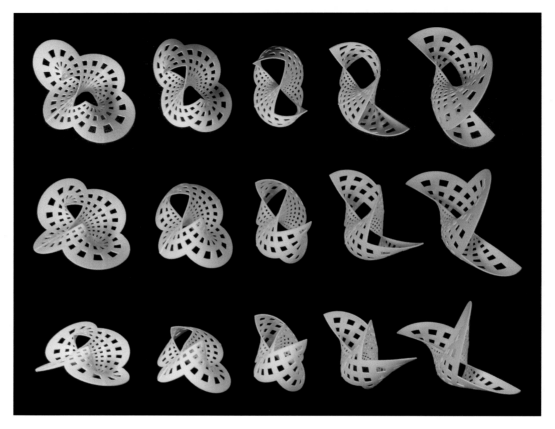

Fig. 6.5. *Top,* Put a dent into the surface. Or, alternatively, take a bite (*red*) out of the side.

Fig. 6.6. *Middle,* Three views of "flipping" the surface.

Fig. 6.7. *Bottom,* Shrink away some of the surface (*red*) and fill in the hole (*green*) to get the usual Möbius strip.

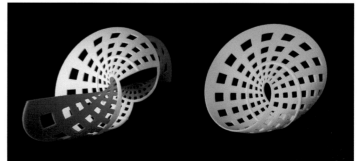

Our mystery surface is a Möbius strip with a hole punched in it. This takes a few steps to see. First, let's push a little dent into the surface at the circular boundary loop (see fig. 6.5). Then, we "flip" the surface around, as in fig. 6.6. The circular boundary loop becomes a straight line, and the other boundary loop becomes a square. Finally, in fig. 6.7, we can shrink the other boundary loop inward and then patching in the square hole gets us the usual picture of the Möbius strip. Apart from the last patching move, everything else we did just distorted the surface. These distortions don't change the intrinsic topology of the surface, just what it looks like in three-dimensional space. So the mystery surface we started with must have been a Möbius strip with a hole.

As with knots (viewed extrinsically, sitting in three-dimensional space), very different-looking pictures of surfaces can really be the same thing. At present, knot theory doesn't have an efficient way to tell whether a picture of a knot is the unknot, let alone whether two pictures of knots are really the same knot. However, there are many partial and heuristic results, and the general problem is known to be solvable in principle (albeit incredibly inefficiently).

Amazingly, there *is* an easy way to tell whether two pictures of surfaces (viewed intrinsically, from the shortsighted ant's perspective) are the same, using the *Euler characteristic* of a surface.

The Euler Characteristic

Let's start with a seemingly innocuous observation: that every polygon has the same number of edges as vertices (see fig. 2.2). This is pretty clear, but as a warm-up, let's really convince ourselves that it is true: We can pair up each edge with the vertex at, say, the clockwise end of it. Each edge gets paired with a vertex and each vertex gets paired with an edge, so there have to be the same number of them. Next, we can think of the vertices and edges of a polygon as really being a tiling of the circle, by radially projecting our polygon to make a "beach ball polygon" on a circle centered on the polygon. This is just like how we made a beach ball cube in fig. 3.10, but one dimension down. One-dimensional tilings are pretty simple: the circle is tiled by some number of edges, and the edges meet their neighboring edges at the vertices to either side of them.

The *Euler characteristic* of such a tiling of the circle (named after the Swiss mathematician Leonhard Euler), is the number of vertices minus the number of edges. Now, given our observation about the numbers of edges and vertices of a polygon, this isn't so interesting. We always get zero, no matter which tiling we look at. But seen another way, this *is* interesting. Because we get the same answer no matter which tiling of the circle we choose, the Euler characteristic is really a feature of the circle *itself*, not of any particular tiling of it. So we can say that *the Euler characteristic of the circle is zero.*

This is a bit of a strange idea. I've only told you how to calculate the Euler characteristic of a *tiling* of the circle, not the circle itself. But because any tiling you choose agrees with any other tiling for calculating the Euler characteristic, it makes sense to think of the Euler characteristic as belonging to the circle itself, rather than a tiling of it.

Things get more interesting when we go up a dimension, from the one-dimensional circle to look at two-dimensional surfaces. For tilings of surfaces, we have faces as well as edges and vertices. To take account of the faces in the Euler characteristic calculation, we add the number of them onto the end of the calculation. So the Euler characteristic of a tiling of a surface is the number of vertices minus the number of edges plus the number of faces.

What do we get if we start calculating the Euler characteristic for different tilings of surfaces? Let's return to the regular polyhedra. See fig. 2.6. The tetrahedron, for example, has 4 vertices, 6 edges, and 4 faces, while the cube has 8 vertices, 12 edges, and 6 faces.

For the tetrahedron, then, the Euler characteristic is 4 − 6 + 4 = 2. For the cube, we get 8 − 12 + 6 = 2. The octahedron gives 6 − 12 + 8 = 2. The dodecahedron gives 20 − 30 + 12 = 2. The icosahedron gives 12 − 30 + 20 = 2. As with tilings of the circle, we always seem to get the same answer. Let's check a few more, such as the archimedean solids (see fig. 2.9). The truncated tetrahedron has 4 triangular faces and 4 hexagonal faces, 18 edges, and 12 vertices. Let's count all of the faces together, and we get 12 − 18 + 8 = 2. The cuboctahedron gives 12 − 24 + 14 = 2. The truncated cube gives 12 − 36 + 14 = 2. I'll leave you to check that the Euler characteristic is always 2.

Or is it? We have seen a tiling that gives a different answer. Take a look at the torus on the left in fig. 5.5. Here, the torus is tiled with a 10-by-10 grid of squares. Counting how many faces, edges, and vertices there are here could take a while. It might be better to think about the unrolled version on the right in fig. 5.5. There are 100 squares. We can count the vertices by matching up the squares with the vertices at their lower left corners. For every square, there is a vertex, and we get all of the vertices this way. So there has to be the same number of vertices as squares: 100. Similarly, every square has an edge to its left and an edge below it, and we get all of the edges this way, so there must be 200 edges: 100 − 200 + 100 = 0. So this time the Euler characteristic is 0, not 2.

There are quite a few other tilings of surfaces in this book. If you're feeling brave, you can try to calculate their Euler characteristics and see whether you get two or something else. But to save you some time, as with tilings of the circle, what you get only depends on the tiled surface. The Euler characteristic is a feature of the surface itself, rather than a tiling of it. The Euler characteristic of the sphere is two. All of the polyhedra are really tilings of the sphere, so you always get two when you calculate vertices minus edges plus faces. The Euler characteristic of the torus is zero. You always get zero when you calculate vertices minus edges plus faces of a tiling of it, as we did for fig. 5.5. The same is true in fig. 5.11; think of it as another tiling of the torus with squares.

Because the Euler characteristic of a surface is the same no matter what tiling we use, we might as well use a nice simple tiling to calculate it. Instead of using a 10-by-10 grid of squares to tile the torus, we could use a single (flexible) square. We can see this in fig. 5.5 again. We start on the right with a single big square with four edges and four vertices. When we roll the square into a tube, the two blue edges become one edge and the four vertices become two. When we roll the tube into the torus, the two green edges become one edge, and the remaining two vertices become one. For this tiling of the torus, there is one vertex, two edges, and one face: $1 - 2 + 1 = 0$.

If you play the same kind of game with the other surfaces we have seen, you'll get the Euler characteristics listed in table 6.1.

There are some beautiful patterns in here:

1. Adding a handle (increasing the genus by one) decreases the Euler characteristic by two; therefore, the Euler characteristic goes down by two as we move from the sphere to the torus to the genus two surface and so on.
2. Punching a hole in a surface decreases the Euler characteristic by one. The Euler characteristic goes down by one each time we move from the sphere to the disk to the annulus.
3. If you have two surfaces with boundary loops and you glue their boundary loops together, you get the Euler characteristic of the new surface by adding together the Euler characteristics of the two surfaces you started with. For example, if you glue two disks together along their boundaries, you get a sphere (think about the northern and southern hemispheres—both are disks—glued to each other along the equator).
4. Similarly, if a single surface has two boundary loops, then you can make a new surface by gluing

Table 6.1 Properties of a selection of surfaces

Surface	Euler characteristic	No. sides	No. boundary loops
Sphere	2	2	0
Disk	1	2	1
Torus	0	2	0
Annulus	0	2	2
Klein bottle	0	1	0
Möbius strip	0	1	1
Möbius strip with a hole	−1	1	2
Genus two surface	−2	2	0
Genus three surface	−4	2	0

those loops together, and the Euler characteristic of the new surface is the same as for the old surface. For example, in fig. 5.5, we made a torus by gluing the two boundary loops of a tube, which is the same as the annulus. Both the annulus and the torus have Euler characteristic zero. Rather than rolling up a square to make an annulus, we could also start with a sphere and punch two holes. If we then glue the holes onto each other, then the whole process of punching two holes and then gluing is the same as adding a handle. So this is another way to see that the Euler characteristic goes down by two each time you add a handle.

There are many relations between different kinds of surfaces, and for each surface, we can write down three numbers: (1) the Euler characteristic, (2) the number of sides, and (3) the number of boundary loops. Now, earlier in this chapter, I promised you an easy way to tell if two surfaces are the same. Amazingly, it turns out that these three numbers do the job. If you find two surfaces that have the same values for these three numbers, then (viewed intrinsically) they are actually the same surface.

A *Seifert surface* for a knot or link is a (two-sided) surface sitting in space, whose boundary loop or loops make up the knot or link. It is often easier to think of a knot or a link as being the boundary of a two-dimensional surface because two-dimensional surfaces are usually easier to deal with than the full three-dimensional way in which a knot or link sits in space. Seifert surfaces are named after German mathematician Herbert Karl Johannes Seifert.

The sets of boundary loops for the surfaces in figs. 6.10 and 6.11 are both three unknots. But as you might guess from the figure captions, for the first these are arranged to form the (3,3) torus link, and for the second the borromean rings.

For example, let's look at the mystery surface in fig. 6.4. Once we know that its Euler characteristic is minus one, it has one side and two boundary components, we know that it must be the Möbius strip with one hole punched out of it, there are no other options.

There is one more part to this story. We can write down the complete list of all possible surfaces as follows: We can make any two-sided surface by starting with a sphere, adding handles, and punching holes. Each added handle reduces the Euler characteristic by two. Each punched hole adds a boundary loop and reduces the Euler characteristic by one. Neither of these moves changes the number of sides the surface has, so it will still have two sides at the end.

We can make any one-sided surface by starting with a sphere, punching holes, and gluing Möbius strips along their boundaries on to some of the holes. Again, each hole adds a boundary loop and reduces the Euler characteristic by one. Gluing a Möbius strip onto a boundary loop doesn't change the Euler characteristic but removes a boundary loop and makes the surface one sided.

And that's it. So, we can identify any surface as being one of these possibilities by counting (to get

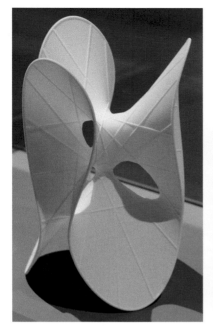

the Euler characteristic and the number of boundary loops) and painting (to see how many sides there are). This whole story is called the *classification theorem for surfaces*: We have a list of all of the possible surfaces, and if someone shows us a surface, we know how to classify it. We know how to figure out which one of the surfaces on our list it is.

If you'd like to try this out for yourself, figs. 6.8 through 6.11 show some complicated-looking surfaces. I'll leave you to figure out what these surfaces are. (The answers are given in appendix A.)

Fig. 6.8. *Left, Clebsch diagonal surface*, by Oliver Labs.

Fig. 6.9. *Right, Minimal monkey trefoil*, by Carlo Séquin.

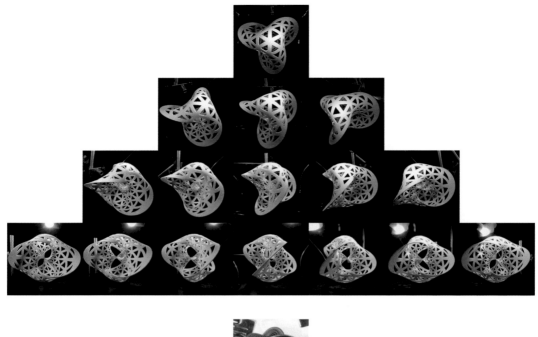

Fig. 6.10. *Top, (3,3) torus link Seifert surface*, by Saul Schleimer and Henry Segerman.

Fig. 6.11. *Bottom, Borromean Rings Seifert Surface*, by Bathsheba Grossman.

The Best Geometry

In this chapter, we have been thinking about surfaces as topological things. But I needed to choose some actual geometric shapes to 3D print. What is the best shape—the best geometry? Let's look at the torus. We already saw many possible shapes for it in fig. 5.4, from a coffee mug–shaped torus to a doughnut-shaped torus, and it's easy to imagine an enormous variety of other possible shapes of torus. But just out of the tori in fig. 5.4, which is the best?

Of course, I have to tell you what I mean by "the best." If, for example, I want my torus to hold coffee and aid in drinking it, then the doughnut-like tori leave something to be desired. But for illustrating the mathematical object, the doughnut-shaped torus is the most symmetric, the most uniform of the ones in fig. 5.4. The coffee-containing shape, although of great interest to mathematicians in its application involving coffee, has a special shape that doesn't really have much to do with the torus as a topological thing. So it is better to get rid of the distraction of the asymmetrical shape.

Can we do any better than the doughnut shape? Is there a shape that is even more symmetrical? The doughnut torus has a rotational axis of symmetry, but not every point of it is the same as every other. In fig. 4.4, you can see the curvature is negative (blue) on the inside of the doughnut hole and positive (red) on the outside. Is there a torus with the same curvature everywhere?

Fig. 5.5 gives a hint. We can tile the torus with a grid of squares, and when we cut and unroll the torus, we get a *flat* grid of squares. We didn't cut the torus in some special way to do this. If in fig. 5.5, you cut along some other green circle that goes through the hole and some other blue circle that goes around the hole, you'll get the same 10-by-10 grid of squares.

At every point, the square grid tiling on the torus looks the same as the square grid tiling of the euclidean plane, with Schläfli symbol {4,4}. The tiling seems to be saying that the torus should act like the euclidean plane. Let's look at this more closely. In fact, let's pretend once again that you are a shortsighted ant. (This means that you have to look at things closely, you have no choice.) This time the torus is tiled by flat squares (see fig. 5.5, right).

Suppose you start in the middle of the grid, far away from the blue and green sides. Suppose you walk one square north, toward the green edge at the top. I don't want you to see the ugly green and blue edges, but luckily for me, you are very shortsighted and you can't see them yet. But you've gone north one square, and if you kept going in that direction, you would run into the green edge. So what I can do is take the southernmost row of squares, move them up, and connect them along the northernmost edge of the grid. You are now back in the middle of the grid. I can do the same if you walk south, west, or east. As you wander around the torus, I can keep moving squares around so that you never see the edge of the grid. Any marks you make on the surface will get moved around with the squares, and when you come back to visit those marks again, everything will be as you left it. As far as you would be able to tell, you're just walking around on the torus, except that the torus is always tiled with flat squares, as if they are on the euclidean plane. So for you, the shortsighted ant, the geometry of the torus as you experience it has zero curvature everywhere. It looks exactly the same as the euclidean plane everywhere. *This*, the *flat torus*, is the best, most symmetric geometry on the torus: every point is the same as every other point.

Great, *intrinsically* the torus should have zero curvature everywhere. But can this be done extrinsically? That is, is there a way to squish a torus around in three-dimensional space so that it has zero curvature everywhere?

Well, it isn't possible to do this if you want the surface to be perfectly smooth. However, it *is* possible to do things less smoothly, for example, if we make the torus out of flat pieces hinged together. See fig. 6.12. This is a torus sitting in space, not crashing through

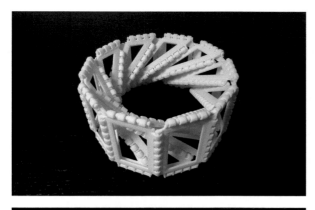

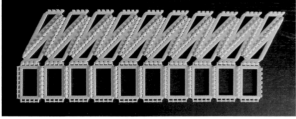

Fig. 6.12. A torus can be made out of a flat tiling of polygons.

or folded flat onto itself. If we cut it in the right places, it can be unrolled out onto the flat, euclidean plane. Does this count as a flat torus though? Would a short-sighted ant be able to tell the difference between crawling around on the plane and on this torus? Well, it depends on what the rules are. If the ant is on a face of this tiling of the torus, then because the face is a piece of the euclidean plane, it wouldn't be able to tell the difference no matter what the rules are. What about on an edge where the faces hinge together? If the ant could tell that the surface is bent along the hinge, then obviously it could tell that it isn't on the flat torus. But if all the ant could do was measure lengths of lines along the surface and angles between those lines, then it wouldn't be able to tell that it wasn't on the plane. This is because we can flatten the hinge out onto the plane, and lengths and angles measured along the surface don't change. So anything the ant does on the folded-up surface is the same as what it would do on the flattened-out surface. The same thing is also true around a vertex. It can also be flattened out onto the plane, because the angle defect at the vertex is zero. At least with this choice of rules, this really is a flat torus.

We don't get a *square* flat torus though, like the one you were crawling around on before in fig. 5.5. We get some other shape that tiles the plane. The paths that go around and through the torus are not the same length, nor are they at right angles to each other, like they are for the square flat torus. Strictly speaking, then, there is more than one "best" geometry for the torus. What I really mean by "best" is *uniform*; that intrinsically, the geometry looks the same everywhere (therefore, the curvature is the same everywhere). This hinged flat torus and the square flat torus are different geometries for the torus, both of them uniform.

What about uniform geometry for other surfaces? The sphere is easy. The usual geometric shape for the sphere (as opposed to some squished version of it), looks the same at every point. The Klein bottle is very similar to the torus in terms of intrinsic geometry. You can cut it along its "neck" and get a cylinder, as with the torus. If you cut it open, you get a square. In the same way as we did for the torus, a shortsighted ant walking around on a square tiling of the Klein bottle would see every point of it as looking like a point of the euclidean plane (the choreography required in moving tiles around to trick the ant would be a bit different; how would it change?).

What about higher genus surfaces? It turns out that a uniform geometry for any of the higher genus surfaces has to be hyperbolic. I'll just show you one example: the genus three surface. See fig. 6.13. To be clear, I'm thinking of the surface in the figure as having a pattern of "filled" triangles (made of plastic) and "empty" triangles (windows that you can see through). I'm not thinking of the empty triangles as holes in the surface.

Well. First of all, it isn't so obvious that this really is the genus three surface. There are a couple of strategies for checking this. We can either try to imagine deforming it in space until it looks like the genus three surface in fig. 6.2, or we can work out the Euler characteristic, the number of sides and loops, and check that we get the same numbers as in table 6.1. The first

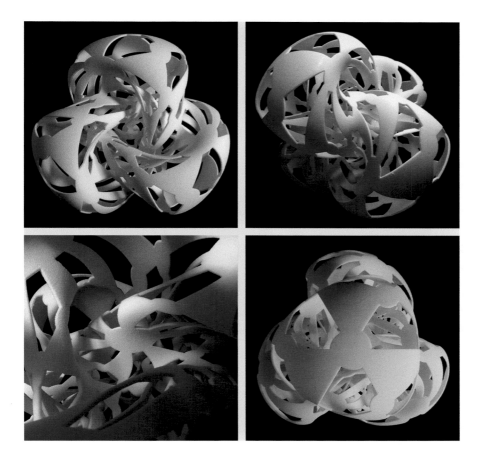

strategy isn't so bad: (see details in appendix A for fig. 6.14). For the second strategy, the number of boundary loops in the surface in fig. 6.13 is obviously zero. Also the surface has an inside and an outside, so it has two sides. Finally, we can calculate the Euler characteristic. We could use the tiling that's already on the surface, although it would be something of an ordeal to count the number of faces (all of which are triangles), edges, and vertices. Easier than that would be to cut the surface up in a different way (see fig. 6.14).

Our surface can be cut up into four pieces, each of which is a sphere with three holes punched out of it. A sphere with three holes punched out of it is called a *pair of pants* (yes, really) because, as far as topologists are concerned, your pants are indistinguishable from a sphere with three holes. Each pair of pants in our surface is connected to its neighbors along its three boundary loops.

Fig. 6.13. A genus three surface, tiled with the (7,3,2) triangle tiling. This is another model that is too complicated to make much sense of even from a grid of photographs—better to look at a three-dimensional model (see 3dprintmath.com).

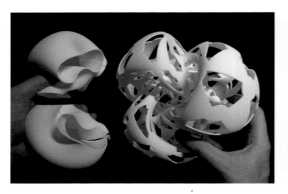
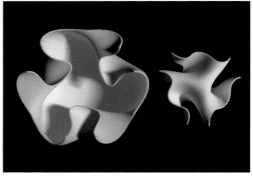

Fig. 6.14. *Left*, the surface is made from four pairs of pants. *Right*, each pair of pants can be made from two (very floppy) hexagons.

A pair of pants can be made from two hexagons, so there are a total of eight hexagonal faces. To count the edges, each hexagon has 6 edges, so there would be $8 \times 6 = 48$ edges, but we are double counting (each edge is adjacent to two hexagons) so there are really 24 edges. Finally, for the vertices, each hexagon has 6 vertices, so there would be $8 \times 6 = 48$ vertices, but we are quadruple counting them (each vertex is adjacent to four hexagons), so there are 12 vertices. The Euler characteristic is $12 - 24 + 8 = -4$. Alternatively, we could get the same answer using the fact that our surface can be made by gluing together four pairs of pants (each of which has Euler characteristic -1), together with the fact that gluing together surfaces along boundary loops adds their Euler characteristics. In any case, everything matches up with table 6.1, and our surface is the genus three surface.

Next, what about uniform geometry on the surface? Just as we did for the torus, we can understand this by looking at the tiling. The tiling in fig. 6.13 is the (7,3,2) triangle tiling, as in fig. 4.12. The tiling is distorted on our surface, as it is for the tiling of the torus in fig. 5.5. In the same way as we did for the torus and the square grid, you can imagine being a shortsighted

The surface in fig. 6.13 is a projection from a higher-dimensional space of something called the *Klein quartic*. This is a fantastically symmetrical surface, named after the same Klein who came up with the bottle. As I've printed this surface, it has 332 symmetry. If you go back to fig. 1.29, you can count that the 332 comma symmetry sphere has 12 commas, corresponding to 12 rotation symmetries. The most symmetric tiling on the sphere is the (5,3,2) triangle tiling, with 60 rotations, or a total of 120 symmetries, including reflections. For the true Klein quartic (rather than a projection of it into three-dimensional space), there is a symmetry that takes *any* triangle of its tiling to any other, for 168 "rotations," or a total of 336 symmetries, including reflections.

ant wandering around on the surface, but thinking that you are wandering around on the (7,3,2) triangle tiling of the hyperbolic plane. This gives a uniform geometry for the genus three surface, with the same negative curvature everywhere.

It turns out that the same kind of thing works for all surfaces with genus bigger than one, and so all of these have hyperbolic uniform geometries, although it isn't always possible to tile these surfaces with the (7,3,2) triangle tiling.

The Gauss-Bonnet Theorem

Looking only at the two-sided surfaces with no boundary loops, uniform geometry for the sphere is the usual geometry on the sphere; for the torus it is the euclidean plane, and for higher genus surfaces, it is the hyperbolic plane. Or, said another way, the sign of the Euler characteristic (+2 for the sphere, 0 for the torus, −2 for the genus two surface, etc.) is the same as the sign of the curvature (positive for the sphere, zero for the torus, negative for higher genus surfaces). This isn't a coincidence. There is an amazing result named after Carl Friedrich Gauss and Pierre Ossian Bonnet that explains this pattern.

The Gauss-Bonnet theorem says that if you add up the curvature over a surface without boundary loops, you get the Euler characteristic times 2π. (By "add up" I really mean "integrate." I won't get into the calculus details here; but roughly speaking, integrating something over a surface means adding up the value of that thing over many very small patches of surface.)

For example, look back at fig. 4.4 again. The torus there has positive curvature (*red*) on the outside and negative curvature (*blue*) on the inside. According to the Gauss-Bonnet theorem, if you add up all of the curvature, you will get precisely 2π times the Euler characteristic of the torus, which is zero. The large area of positive curvature on the outside is *precisely* balanced by the smaller area of more strongly negative curvature. And the Gauss-Bonnet theorem doesn't just apply to this very nice torus, it applies to every possible geometry on the torus. The same is true for all of the more-or-less coffee mug–shaped tori in fig. 5.4. The big dent in the surface where the coffee is supposed to go will have some positive curvature in it, because it is shaped approximately like part of a sphere. This positive curvature here is balanced out by some negative curvature somewhere else.

This is a truly amazing result: the Euler characteristic only depends on the topology of the surface, but the curvature depends on the geometry, which could be almost anything. Somehow, if you add up all of the curvature, coming from whatever geometry you put on the surface, the total doesn't depend on the geometry of the surface at all—only the topology.

The Gauss-Bonnet theorem is easiest to check for the uniform geometry on the torus. The euclidean plane has zero curvature everywhere, so if you add zero up over the torus, you get back the correct answer of zero. This also tells us that uniform geometry on the torus must be the euclidean plane. It cannot be the sphere or the hyperbolic plane. If we could put spherical geometry on the torus, then adding up positive curvature everywhere would give a positive answer, but Gauss-Bonnet tells us that we have to get zero. The same argument works for hyperbolic geometry. Adding up negative curvature everywhere would give a negative answer. The same kind of argument shows that the sphere can only have spherical uniform geometry, and higher genus surfaces can only have hyperbolic uniform geometry.

We can now return to a mystery I left you with way back at the end of chapter 2, in table 2.1. Why was it that the sum of the angle defects for every polyhedron we looked at gave 720 degrees? Well, 720 degrees is otherwise known as 2 × 360 degrees or, expressed in radians, 2 × 2π, which is the same as the Euler characteristic of the sphere times 2π. This comes from a version of the Gauss-Bonnet theorem with the angle defect at the vertices standing in for curvature. In the same way, adding up the angle defects for all vertices of, say, the polyhedral torus on the left of fig. 5.5, you'll get zero. It's easier to see that this will work for the flat, hinged torus in fig. 6.12. The angle defect at every vertex is zero because the tiling of polygons can be flattened out onto the table.

Uniform Geometry for Surfaces with Boundary

What about uniform geometry for surfaces with boundary? There is a problem here: We want every point of the surface to look the same as every other point, but how can this be if you can walk up to the boundary and fall off? The boundary of a surface looks very different from some other point in the middle of it. The solution is to make the boundary infinitely far away. For example, we could view the annulus as an infinitely long cylinder. Then every point looks like a point of the euclidean plane. There is no boundary to stand on. What about a pair of pants?

Fig. 6.15 shows one way to put uniform geometry on a pair of pants: take two hyperbolic triangles from the {3,∞} Schläfli tiling, as in fig. 4.16, and glue them together along all three edges. This is like making a triangular pillow, except that the corners of the pillow are infinitely far away—so they aren't actually part of the surface. This is where the three holes go. As with

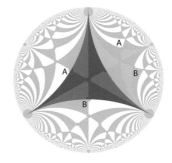

Fig. 6.15. Glue the A edges to each other and glue the B edges to each other to make a pair of pants from these two triangles of the {3,∞} Schläfli symbol tiling.

coffee mugs, the "best" geometry may not actually be that good for items of clothing, unless your legs happen to be infinitely long.

I should mention that there is also a version of the Gauss-Bonnet theorem for surfaces with boundary loops, but the connection to the classification theorem is a bit messier than the nice, neat story for surfaces without boundary loops I've outlined here.

Uniform Geometry in Three Dimensions

Many of the things we have been talking about in this chapter carry over when we increase the dimension, from two-dimensional surfaces to three-dimensional *manifolds*. When you zoom in on a small part of a surface, it looks like a two-dimensional plane. Similarly, a three-dimensional manifold (or *three-manifold*) is a thing that looks like three-dimensional space when you zoom in on a small part of it. Three-dimensional space itself is an example, as is the three-sphere from chapter 3 (see fig. 3.15).

Fig. 6.16 shows another example of a three-manifold—the *complement* of a knot. This is what you get when you start with a solid ball, and you tunnel out the knot. Perhaps you could train a worm to eat through the ball, following the path of the knot. In the case of fig. 6.16, we tunnel out the figure-eight knot. This is closely related to the print shown in fig. 5.11: in fact, they are stereographic projections from the three-sphere to three-dimensional space of the same thing—a tube around the knot. In fig. 5.11, the light source for stereographic projection is quite far away from the knot, while in fig. 6.16, it is on the knot itself. This means that a point on the knot gets projected to infinity in three-dimensional space. We are on the inside of the tube here, and the complement of the knot is the tunneled-out ball in front of us. In fig. 5.11, we are outside of the tube, and we are inside the complement of the knot, along with the rest of the universe. In both cases, the manifold is the same, we are just looking at it in two different ways.

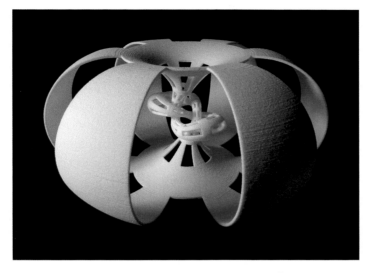

Fig. 6.16. The figure-eight knot complement.

This knot complement is an example of a three-manifold with boundary. Two-dimensional surfaces can have one-dimensional boundary loops. Likewise, three-dimensional manifolds can have two-dimensional boundary surfaces, in this case a boundary torus: the tube around where the knot was before our trained worm tunneled it out.

How are things different for three-manifolds? Can we classify three-manifolds in the same way that we classified surfaces? One of the important tools for classifying surfaces was the Euler characteristic. This still makes sense for three-manifolds. To go from tilings of the circle to tilings of surfaces, we added the number of two-dimensional faces to the Euler characteristic calculation. To go from tilings of surfaces to tilings of three-manifolds, the right modification to make is to subtract the number of three-dimensional cells. So, for example, let's calculate the tiling of the three-sphere given by the hypercube. See table 3.1. We get vertices – edges + faces – cells = 16 – 32 + 24 – 8 = 0.

Unfortunately, the Euler characteristic isn't any-where near as useful as it was for surfaces. It turns out that if a three-manifold doesn't have any boundary surfaces, then its Euler characteristic is always zero. So this isn't going to help much in terms of identifying three-manifolds.

There *is* an analogous classification theorem for three-manifolds, but it is a much more complicated story than the one for surfaces. This is the *geometrization theorem*, which was conjectured by the American mathematician William Thurston in 1982 and finally proved in 2003 by the Russian mathematician Grigori Perelman, building on work by another American mathematician, Richard Hamilton. This theorem also settled the Poincaré conjecture, a closely related question that had stood since 1904.

Using the Gauss-Bonnet theorem, we can think of the classification of surfaces (without boundary loops) as really talking about uniform geometry on a surface. It is either spherical, euclidean, or hyperbolic, depending on whether the Euler characteristic is greater than, equal to, or less than zero. For three-manifolds, we get more than three geometries: the geometrization theorem gives us eight. Three of them are relatively familiar: the three-sphere, ordinary three-dimensional euclidean space, and three-dimensional hyperbolic space, as we saw at the end of chapter 4. The other five geometries are various combinations of lower-dimensional geometries, for example, a geometry that is hyperbolic in two dimensions but euclidean in the other one. We also don't get that every three-manifold has one of these eight geometries. Instead, we may first have to cut the three-manifold up into pieces along spheres and tori and then the pieces have one of the eight geometries.

A Hyperbolic Three-Manifold

Let me show you one example of a three-manifold with hyperbolic uniform geometry (no cutting up as in the geometrization theorem is needed beforehand). This manifold is the complement of the figure-eight knot, as we saw in fig. 6.16. The hyperbolic geometry on this knot complement was first discovered by Robert Riley in 1974. The description I'll give you is due to William Thurston.

We are going to do this in the same way as we did for surfaces. We want to find a tiling of the complement of the figure-eight knot, which from the short-sighted ant's point of view, looks the same as a tiling of three-dimensional hyperbolic space. (Previously, our ant was crawling around on a surface; now it had better be a flying ant so that it can move around in three dimensions.)

The tiling of three-dimensional hyperbolic space we want to use is the {3,3,6} Schläfli symbol tiling, as seen in fig. 4.23. This is a tiling of hyperbolic space by tetrahedra, with six meeting at every edge. Just like the two triangles we used to tile a pair of pants in fig. 6.15, our tetrahedra have vertices that are infinitely far away. However, it will be easier to see the tiling if we use truncated tetrahedra (see fig. 2.9, no. 1). This is analogous to tiling the pair of pants with two hexagons. Each hexagon has three "long" edges that are glued to the other hexagon and three "short" edges that are on the boundary loops. Similarly, each truncated tetrahedron has four "big" hexagonal faces that we will glue to other truncated tetrahedra and four "small" triangular faces that are on (and, in fact, tile) the boundary surface. When we put uniform (two-dimensional hyperbolic) geometry on the two hexagon tiling of the pair of pants, the short edges shrink away to nothing, and the hexagons turn into triangles that are missing their vertices because they have vanished off to infinity. Likewise, when we put uniform (three-dimensional hyperbolic) geometry on our tiling of the figure-eight knot complement, the small triangles shrink away to nothing, and the truncated tetrahedra turn into ordinary tetrahedra that are missing their vertices.

All I have to do is show you how to cut up the complement of the figure-eight knot into truncated tetrahedra. The triangular faces need to go on the boundary surface—the torus where the knot was tunneled

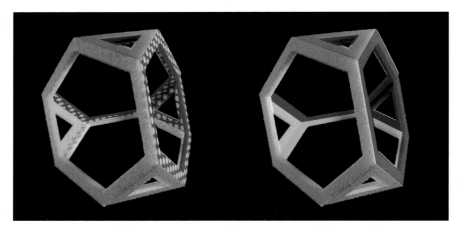

Fig. 6.17. Two truncated tetra-hedra glue together to make up the complement of the figure-eight knot. These are colored versions of the model in fig. 2.9, no. 1.

out. And the edges of the tiling that go through the inside of the three-manifold (as opposed to along the boundary torus) should have six truncated tetrahedra around them. This is because these are the edges that are still there after we shrink away the triangular faces, and so they need to turn into the edges of the {3,3,6} Schläfli symbol tiling, each of which has six tetrahedra around them. It seems hard to believe, but in fact this can be done with only two tetrahedra. See fig. 6.17.

But hold on a minute. How can we fit six truncated tetrahedra around each edge if there are only two of them in the tiling? This isn't so crazy. In this chapter, we thought of fig. 5.5 as showing a tiling of the torus by a single square. There, the single vertex looks like it has four squares arranged around it. The square is so distorted that its four corners end up at the same vertex. Similarly, when we glue our truncated tetrahedra together to make the complement of the figure-eight knot, they get so distorted that three of the edges of each truncated tetrahedron get glued together into one edge. If you count up the numbers, this means that the tiling has only two edges that go through the inside of the three-manifold. Again, this isn't so crazy. The tiling of the torus in fig. 5.5 has only one vertex.

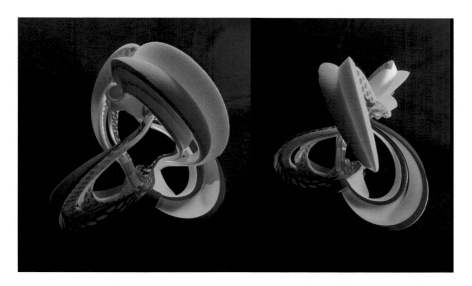

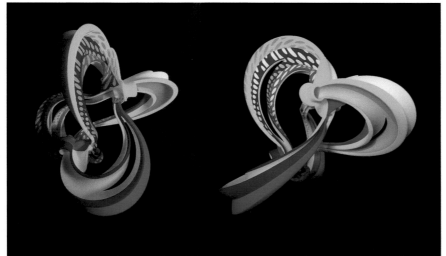

This tiling of the complement of the figure-eight knot isn't easy to see; in fact, this is the hardest thing to visualize in the book. Even with a 3D printed model, or one that you can rotate around on screen, this is difficult. But here we go (see fig. 6.18). This is colored in the same way as the truncated tetrahedra in fig. 6.17. Let's pause for a second to see how the two tetrahedra are colored. The hexagonal faces are colored white, red, green, and blue, and the triangular faces yellow, cyan, magenta, and black. One of the two tetrahedra also has gray polka dots. These two tetrahedra are glued to each other along their matching color hexagonal faces to make fig. 6.18.

Fig. 6.18. Triangulation of the figure eight-knot complement.

In fig. 6.18, the figure-eight knot is arranged in space symmetrically, in the same way as in fig. 5.11. The polka-dot truncated tetrahedra is on the "inside," and the plain truncated tetrahedron is on the "outside." We are viewing the faces of the tetrahedra from within the outside truncated tetrahedron.

First, let's see where the eight triangular faces of the two truncated tetrahedra are. These tile the boundary torus around where the knot was before we tunneled it out. The triangles are very stretched out—each triangle has one edge that wraps around the tube, while the other two reach a full quarter of the way along the knot. The cyan triangle is visible in the lower images in fig. 6.18, wrapping around the rightmost part of the knot.

The hexagonal faces of the two tetrahedra are glued to each other, matching colors, and separate the inside truncated tetrahedron from the outside one. Each hexagon has alternating short and long edges. Here "short" means that it is an edge shared with a triangular face, which goes away when we shrink the triangular faces away. The "long" edges are shared with other hexagons. In fig. 6.18, the short edges go along the knot and in this model are longer than the long edges that go through the three-manifold and jump from one part of the knot to another. One of these long edges is at the top of the model, and the other is at the bottom. Counting the number of hexagon edges that meet at a long edge, we get six, which is the number we expect. It's the same as the number of tetrahedra meeting around that edge. The top edge sees the green hexagon twice, the white hexagon twice, and the red and blue hexagons once each. The bottom edge sees the red hexagon twice, the blue hexagon twice, and the green and white hexagons once each.

Well, and that's it. It is very difficult to see each distorted, truncated tetrahedron all at once in fig. 6.18, but it isn't so hard to follow its triangles and hexagons around, matching them up with the undistorted versions in fig. 6.17. This really is a tiling of the complement of the figure-eight knot by these two truncated tetrahedra. When we shrink away the triangles, we get a tiling by tetrahedra missing their vertices, which from the shortsighted flying ant's point of view is the same as the {3,3,6} Schläfli symbol tiling of three-dimensional hyperbolic space (see fig. 4.23). As the ant flies around, as far as she can tell, she is in three-dimensional hyperbolic space, and the geometry looks the same at every point.

The uniform geometry for the complement of the figure-eight knot is hyperbolic. What about the complements of other knots? Amazingly, almost all other knot complements also have hyperbolic geometry. I mentioned in chapter 5 that there are just over 1.7 million prime knots with minimal crossing number of 16 at most; of these only 32 are not hyperbolic.

Soap Films and Minimal Surfaces

As with knots, we often think of surfaces as topological objects, and we have to choose some geometric shape for them to actually print. Again, as with knots, there are roughly three different ways of doing this, the by-hand method, the parametric/implicit method, and the iterative method. In this chapter, the prints in figs. 6.2 and 6.12 were made by-hand, fig. 6.11 by an iterative method, and the others mostly using parametric or implicit methods (with some combinations of multiple methods, for example, fig. 6.18 has both parametric and by-hand aspects).

In chapter 5, we saw an example of the iterative method with a physical interpretation: pulling a knot as tight as possible. There is a beautiful analogous situation with surfaces. Rather than pulling a rope tight, we can imagine pulling a surface tight. One physical interpretation for this is what a soap film does.

Take a loop of wire (or 3D printed plastic), dip it into a bowl of soapy water, and pull it out again. If you are lucky, you will see a soap film clinging to the wire loop (see fig. 6.19). If you blow gently on the soap film or slowly wave the loop around in the air, you can

Fig. 6.19. A soap film on a loop.

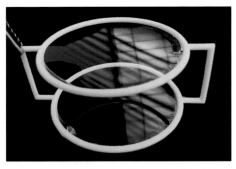

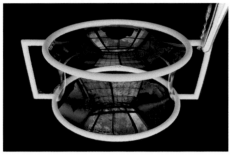

Fig. 6.20. Two different soap films with the same boundary.

make the surface wobble and bulge. But unless you break the film, when you stop blowing or waving it quickly snaps back to the same shape every time.

What is this shape? How can we describe it? Surface tension causes the soap film to pull on itself, reducing its area as much as possible. So our first thought might be that the soap film is a *least area surface*, meaning that it has the least area of all surfaces that have the same boundary loop (or loops) as it does.

However, this isn't quite right. There might be more than one way in which a soap film can cling to its boundary loops (see fig. 6.20). Here, the boundary is two circles, and the soap film could be either two disks or the "waisted cylinder" surface below, which is called the *catenoid*. Depending on how far apart the two circles are, either the two disks or the catenoid have the least possible area. If the two circles are very close together, the catenoid has less area. If they are very far apart, the two disks do better. But wherever the circles are, if the soap film is already in the shape of the surface with more area, it cannot "jump" to find the surface with less area. Any small change in the shape will increase the area, and surface tension doesn't allow that, so the surface is stuck where it is.

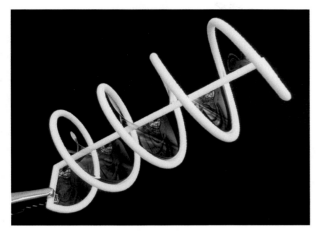

Fig. 6.21. A soap film in the (approximate) shape of a helicoid.

Physics only lets the shape of the soap film change slightly when it tries to reduce its area.

Here is one way to give the definition of a *minimal surface*, a mathematical model of a soap film that takes account of the surface not necessarily being the least possible area. A surface is *minimal* if for every point of the surface there is a small patch around it that is a least area surface.

This means that the surface is made up of lots of little patches, each of which truly has the least area possible given the shape of its boundary, even though the surface as a whole might not minimize area. This also lets us throw away the boundary loops and think about just the surface, because our definition doesn't require that the whole surface has any particular boundary.

In 1760, Joseph-Louis Lagrange first considered the mathematical problem of finding minimal surfaces but was only able to prove that one surface, the plane, is minimal. Soon after, in 1776, Jean Baptiste Meusnier added the catenoid (which we saw in fig. 6.20) and the helicoid (see fig. 6.21). Gradually, over the centuries more and more examples were found.

As the examples got fancier, it became more diffi-

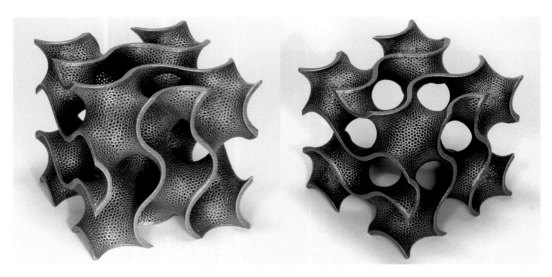

Fig. 6.22. *Gyroid*, by Bathsheba Grossman.

cult to make them as actual real-life soap films, rather than as abstract mathematical soap films. Thankfully, we have 3D printing to show us what they look and feel like. One of the most beautiful minimal surfaces is the *Gyroid*, discovered by Alan Schoen in 1970. Fig. 6.22 shows a chunk of the gyroid, realized as a sculpture by Bathsheba Grossman.

Bathsheba's sculpture shows only part of the gyroid. In fact, the gyroid extends infinitely far in all directions. It is easier to see how this works if we look at only part of Bathsheba's sculpture, a *repeating unit*, shown in fig. 6.23. If we tile ordinary three-dimensional space with cubes in the {4,3,4} Schläfli symbol tiling (as in fig. 4.17) and then put one of these units in each cube, their boundary curves connect up. See fig. 6.24. (Note that this doesn't quite work if you use Bathsheba's larger chunk of the gyroid.) So this unit can be stacked up, repeating in all three dimensions: left/right, forward/backward, and up/down. It glues together to form a single minimal surface, in particular a *triply periodic* minimal surface because it repeats in all three dimensions.

Why don't we see soap films in the shape of the gyroid? It turns out that although each tiny patch of the

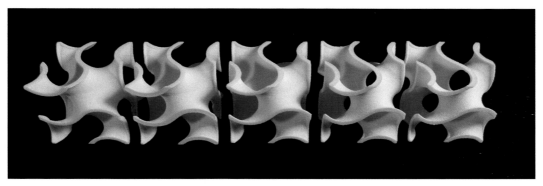

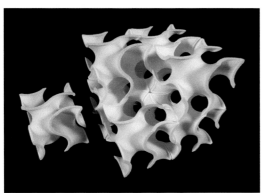

Fig. 6.23. *Top,* Five copies of a repeating unit for the gyroid.

Fig. 6.24. *Bottom,* Eight copies of the repeating unit stack to make a 2 × 2 × 2 cube.

gyroid has the least area possible relative to its boundary, moving all of the patches at once can reduce the area. If we did ever manage to produce a soap film with this shape, it would quickly deform as a whole, collapse, and end up as a much more ordinary soap film surface. Fig. 6.25 shows what happens if you try, using the boundary of the repeating unit as a frame.

Fig. 6.25. Real soap films don't find the gyroid shape.

This story is an example of a common theme in mathematics: The study of minimal surfaces was originally about a physical phenomenon—soap films. But the mathematical model took on a life of its own and grew beyond its original scope to encompass many other strange and wonderful things.

7 Menagerie

Welcome to the end of my book, dear reader. There is much more that I would have liked to say, given more space and time. And there are many more wonderful mathematical ideas to illustrate with 3D prints. I only have space to very briefly show you a few of my favorite prints that I can't cover in more detail.

Fractals

First, Nervous System is a generative design studio that specializes in using computer simulations of natural processes to make art, jewelry, and housewares. Fig. 7.1 shows one of their projects: lamp designs based on a computer simulation of the growth of veins in leaves. Their simulation is based on a model developed by Adam Runions, who is currently a postdoctoral researcher at the Max Planck Institute for Plant Breeding Research. This is an iterative design process quite unlike the others we have looked at. A growing leaf is not trying to minimize the length of rope used in a knot or pull a soap film tight, but it does follow a process that incrementally changes its design over time.

The pattern of veins on a leaf is an example of a *fractal*. Roughly speaking, a fractal is an object that

Fig. 7.1. *Hyphae lamps*, by
Nervous System.

looks similar to itself at many different scales. The
pattern of branching from the thickest vein of a leaf is
approximately repeated on smaller and smaller veins.

Fig. 7.2 shows a much more rigid fractal shape
than Nervous System's Hyphae lamps, a mobile designed by Marco Mahler and me. The veins of leaves
are very similar in structure to the branches of a tree,
although trees differ in that there are usually no loops
in the branches. Like many mobiles, this mobile is
(mathematically) a tree. At the top is the "trunk," a
tetrahedron. The trunk branches into three smaller
tetrahedra hanging from its three lower vertices. Each
of these branches into three smaller tetrahedra and so
on for two further levels.

Well, I say that this is a more rigid shape than the
Hyphae lamps. In one sense, it is obviously more
flexible, because as a mobile it moves around with the
air. However, from the point of view of the mathematical construction, it is much more rigid. The way
it branches is the same everywhere, and the structure
is simple to describe in comparison to the complex,
organic Hyphae lamps.

Another rigid fractal shape comes from a process
that produces a *space-filling curve*. Fig. 7.3 shows

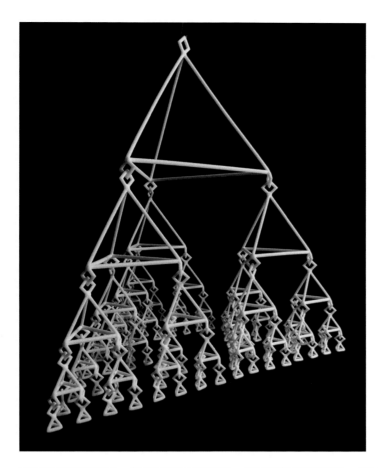

Fig. 7.2. *Ternary tree mobile*, by Marco Mahler and Henry Segerman.

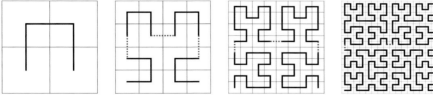

the first few steps in the construction of one of these space-filling curves, the *Hilbert curve*. Subsequent steps in the sequence are more and more squiggly, so that the curve fills up more and more of the square. After infinitely many steps, we get the Hilbert curve. Space-filling curves break the intuitive idea of how many dimensions something has. In one sense, it is still a one-dimensional curve, and yet it goes through every point of the two-dimensional square.

One of the interesting features of this sequence of curves is that from step to step each piece of the curve gets more complicated but doesn't move around that

Fig. 7.3. Four steps in the construction of the Hilbert curve, which was discovered/ invented by the German mathematician David Hilbert. To get from one step to the next, take four copies of the previous curve, scale them down by a factor of two, rotate and position them as shown, and then connect the copies up by adding the dotted red lines.

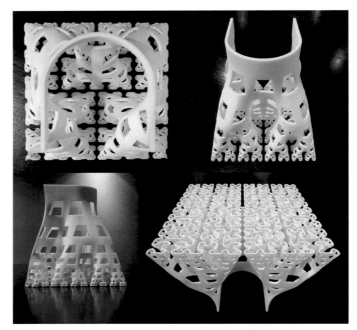

Fig. 7.4. *Developing Hilbert Curve*, by Henry Segerman.

Fig. 7.5. *Hilbert Cube 3D*, by Carlo Séquin.

much. This feature is necessary for it to make sense to talk about the sequence of curves approaching a space-filling curve. If the curves changed wildly from one step to the next, then it wouldn't make sense to talk about what happens after infinitely many steps. The curves have to settle down after a while. Another consequence of this "settling down" feature is that we can smoothly morph between steps of the curve, and the intermediate curves during the morph are nice and well behaved. This is what's going on in fig. 7.4, with the morphing happening vertically as we move down the print.

Going up a dimension in a different way, there are similar constructions of curves that fill the three-dimensional cube instead of the two-dimensional square. Fig. 7.5 shows a sculpture by Carlo Séquin, the third step in the construction of one of these 3D versions of the Hilbert curve.

More Surfaces

In chapter 6, we looked at the problem of making a flat torus in three-dimensional space. Fig. 6.12 showed a way to make this out of hinged flat pieces. This isn't a nice smooth surface because of the fold lines. In the language of calculus, the hinged flat surface is continuous, but it isn't differentiable. There are points on the surface (on the hinges) where the "slope" (really, the tangent plane) of the surface doesn't make sense, because the two faces that meet at the hinge disagree over what the tangent plane should be. It isn't possible to find a perfectly smooth flat torus in three-dimensional space, but it is possible to make a slightly smoother version than the hinged flat torus. Fig. 7.6 shows this very wrinkly torus. In fact, despite all of the wrinkles, this is a *square* flat torus. If we measure distances and angles along the surface, we get exactly the same answers as for the square flat torus we talked about in chapter 6. The paths around and through the torus are the same length, and they are at right angles to one another. The fractal wrinkles are necessary to make this work. We can think of the shape by building it up, starting from the doughnut-shaped torus (as in fig. 4.4) and adding more and more ripples. The first, biggest set of ripples makes the path that goes through the torus longer, to better match with the path that goes around the torus. Subsequent sets of ripples adjust and readjust these lengths, approaching ever closer to the lengths and angles for

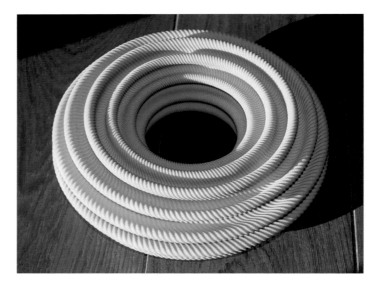

Fig. 7.6. A slightly smoother flat torus, generated by Vincent Borrelli, Saïd Jabrane, Francis Lazarus, and Boris Thibert.

the square flat torus. After infinitely many steps, we get there. Despite having infinitely many ripples upon ripples, the sizes of subsequent sets of ripples shrink fast enough that the result turns out to be slightly smoother than the hinged, flat torus. It has a tangent plane at every point, which varies continuously as we move around the surface.

Also in chapter 6, we looked at the best (meaning uniform) kinds of geometry for various surfaces. We did this by finding nice, regular tilings of those surfaces and took the geometry on the surface from the tiling. Fig. 7.7 comes from thinking about a reverse problem. If you have a surface that already has geometry (in this case, the shape of a bunny), can we put a tiling on it that is "nice" relative to the geometry? In this case, "nice" means that the tiles (faces) all have four sides, most vertices have four faces meeting at them, and the angles at the corners of the faces should be near to 90 degrees. So the tiling should be close to a square grid in most places, varying from this only to take account of the curvature of the surface.

This bunny is tiled by 72 copies of the word "Bunny" (a similar self-referential idea to the print in fig. 1.37). Each copy of the word roughly fits on a

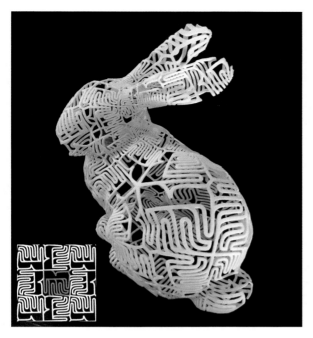

Fig. 7.7. *"Bunny" Bunny*, **by Craig S. Kaplan and Henry Segerman. The pattern on the four-sided tiles is shown in the lower left.**

four-sided tile (in fact, the letters "u-n-n-y" together with a "B" lying on their tops more closely approximate the tile, as shown in the lower left of fig. 7.7). The bunny is topologically a sphere, so its Euler characteristic is two. With this, the fact that each tile (face) has four edges and four vertices, and knowing that there are 72 faces, it's possible to work out that each vertex has, on average, slightly fewer than four faces meeting at it (in fact 4/37 less than four). So we would expect to see lots of vertices with four faces and some with fewer than four. However, there is a vertex at the tip of each of the two ears with only one face next to it (corresponding to very positive curvature) and a few others with three faces. This is balanced out by a number of vertices with more than four faces (roughly corresponding to points with negative curvature). The pattern gets mapped onto each tile in an approximately angle-preserving way. The size of the design varies, but it's mostly undistorted otherwise.

Mechanisms

Oskar van Deventer is a prolific puzzle designer, a member of a community of designers using 3D printing. Many (but by no means all) of his puzzles

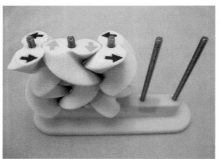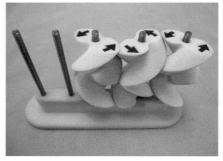

Fig. 7.8. *Top, Over the Top 17 ×
17 × 17,* by Oskar van Deventer.

Fig. 7.9. *Bottom, Magic Gears,* by
Oskar van Deventer.

are "twisty puzzles," inspired by the classic Rubik's cube. The cube has been generalized in an astonishing variety of ways: to all of the other regular polyhedra, many irregular polyhedra, with parts that rotate in many different ways, and so on. Oskar holds the world record for the largest physical Rubik's cube variant, with his *Over the Top 17 × 17 × 17* cube (see fig. 7.8).

Twisty puzzles are connected to some of the topics we have looked at in this book. They are generally made out of polyhedra, and the way they can be twisted is closely related to symmetries of those polyhedra. But there is much more going on in their designs; they are feats of engineering. It isn't obvious how an ordinary 3 × 3 × 3 Rubik's cube holds itself together as it turns, let alone something as complicated as Oskar's *Over the Top 17 × 17 × 17*.

On the subject of surprising mechanisms, fig. 7.9 shows *Magic Gears*, also by Oskar van Deventer. Three strangely shaped gears fit onto axles mounted on a base. On the left, the gears work in the same way as ordinary gears. If you turn the first gear counter-

Fig. 7.10. *Triple gear*, by Saul Schleimer and Henry Segerman.

clockwise, then the second turns clockwise, which makes the third turn counterclockwise. The directions that the gears turn alternate as you go along. On the right, however, if you turn the middle gear upside down, then, paradoxically, all three gears turn in the *same* direction. (The spacing of the axles has to be slightly different as well.)

Finally, also on the subject of unusual gears, fig. 7.10 shows *Triple gear*, another strange mechanism. These three rings make the (3,3) torus link (see fig. 5.7). As I mentioned in chapter 5, the Borromean rings have been used as a symbol of unity, because cutting any of the three loops makes the other two fall apart. This doesn't happen for the (3,3) torus link, but *Triple gear* is unified in a different way. No single gear can turn on its own, but all three rings can turn together.

The Future

I hope that you have enjoyed this brief tour of 3D printable mathematics. There are many more beautiful 3D prints that I could have talked about, and many, many more that I couldn't, because they haven't been designed yet. 3D printing technology and 3D design software are both getting ever more accessible, so if you want to get started on filling in those gaps, that would be great. 3D printing is finding its way into high school and college courses, and the Maker Movement is bringing these tools to everyone. Making a 3D printed model of a mathematical object is an excellent way to learn about it. By the time you're done, you know it very well—you have to in order to be able to tell the computer and the printer what the shape is. So I encourage you to try designing your own. It's difficult for me to recommend a good resource for this, because everything is both new and changing quickly, but you might find it useful to look at the appendix B ("How I Made These Models") for how I do things at the moment. Also take a look around online, there are lots of free resources and programs out there.

Almost all of the 3D models in this book are available to download from the website 3dprintmath.com, and many of these may be good starting points for printing mathematical models on your own printer. Or maybe not. For most of the prints in this book, I used Shapeways, an online 3D printing service, which has high-end 3D printers. The designs in this book may or may not work so well on a home printer. But 3D printing technology is also changing quickly, so printing will get easier and better. My hope is that people will take my 3D designs and modify them to work better with different printing technologies and build on those models to make further visualizations.

So, happy printing, happy mathematics, and I'll see you in the future.

Appendix A: Commentary on Figures

This section provides further details on some of the figures and the 3D prints shown in them. Unless otherwise indicated, all figures, 3D prints, and photographs are by me.

Fig. 1.1: The comma symbol is particularly good for showing symmetries because it is perhaps the simplest symbol with no symmetry itself. It is also good for 3D printing because it has no small features that would be difficult to print. As far as I'm aware, John Conway first used the comma in this way, in an early draft of *The Symmetries of Things*.

Fig. 1.3: Bubble blown by Saul Schleimer.

Fig. 1.7: Paper windmill made by Jil Segerman.

Fig. 1.14: Thanks to the Oklahoma State University Physics and Chemistry Instrument Shop for the construction of the clamp stand, and to Joyce Lucca and Sam Welch of the Institute for Teaching and Learning Excellence for the loan of the turntable.

The first time I tried to make an image with this rig, I lit the object using standing lamps. The problem with this is that the lighting appears to change as you look around the grid of pictures of the object. In fact the lights are fixed, and it is the object that is rotating (relative to the center of the Earth), but the effect is the same. To fix the problem, I needed to have the lights rotate with the object. So I attach clamp lights onto the rig (not shown in the photograph), and turn off all other lights.

Fig. 1.19: This print differs from the later comma symmetry spheres in that it doesn't have small holes at the rotational symmetry points or raised lines at the mirror planes. These would clutter the print. Here I just want to show the commas.

Fig. 1.20: 3D print and photographs by Bathsheba Grossman. Bathsheba made this with the by-hand method, using the 3D design program Rhinoceros, based on a prototype made out of vinyl tubing and plasticene.

Fig. 1.21: I wrote a Python program to combine photographs from the camera into arrays of images as in this figure and the other grids of photographs. There is a difficult issue here in choosing how to orient each photograph before combining them (or, equivalently, choosing the roll angle for each camera in fig. 1.22). Assuming we want these angles to change continuously, no canonical global choice is possible. Such a choice would give a counterexample to the hairy ball theorem. Therefore, for each symmetry type, I had to make a more or less ad hoc choice for the orientations of the photographs in the array. To do this, I chose by hand the orientations of photographs at the corners of the panels (which are really fundamental domains for the symmetry group), retaining symmetry where possible, and then used quaternion interpolation to determine the orientations of the others.

Fig. 1.22: If I positioned cameras around *Soliton* as shown and took photographs, I would get the photos displayed in fig. 1.21. However, the camera rig doesn't have a way to roll the camera around the direction it is looking. To fix this, I rotate the photographs by the appropriate angles in the Python program.

Fig. 1.30: 3D print and photographs by Bathsheba Grossman. *Double Zarf* has symmetry type 223.

Fig. 1.31: 3D print and photographs by Bathsheba Grossman. *Tentacon* has symmetry type 2×.

Fig. 1.32: 3D print and photographs by Bathsheba Grossman. *Metatrino* has symmetry type 432.

Fig. 1.33: 3D print and photograph by George W. Hart (http://georgehart.com). *Six Nested Truncated Cuboctahedra Centerpiece* has symmetry type *432.

Fig. 1.34: 3D print and photograph by George W. Hart (http://georgehart.com). *Solar Centerpiece* has symmetry type 532.

Fig. 1.35: 3D print and photographs by Vladimir Bulatov (http://bulatov.org). *Moebius II* has symmetry type 225.

Fig. 1.36: 3D print and photographs by Vladimir Bulatov (http://bulatov.org). *Rhombic Triacontahedron IV* has symmetry type 532.

Fig. 1.37: *"Sphere" Sphere* has symmetry type 22(10).

Fig. 2.6: This style of representation of polyhedra (hollow with windows cut into the faces) seems to have been first used by Leonardo da Vinci in his illustrations for Luca Pacioli's *De Divina Proportione* (1498). In these models, the "shell" is constant thickness, and the window frames are constant width.

Fig. 2.7: 3D prints by Laura Taalman. These Poly-Snap tiles were made using a Makerbot Replicator 2.

Fig. 3.4: Perhaps our two-dimensional friend is a flatlander, from Edwin A. Abbott's 1884 novella *Flatland: A Romance of Many Dimensions*.

Fig. 3.5: 3D print by Bathsheba Grossman. For both this model and the one shown in fig. 3.9, Bathsheba implemented George Hart's algorithm for thickening edges (see http://www.george hart.com/solid-edge/solid-edge.html).

Fig. 3.6: The first steps in this sequence give us a hint for what the coordinates of the vertices of the hypercube should be. We can think of the vertices of the line segment as being at positions 0 and 1 on the line. When we take a copy of the line segment and move it to the side, we add a dimension to the vector that describes the positions of its vertices. So the square has four corners: at (0,0), (1,0) (the two vertices of the original line segment), (0,1), and (1,1) (the vertices of the copy). Every time we go up a dimension, we add a number at the end of each vector, which can be either 0 or 1, doubling the number of vertices. For example, the four-dimensional hypercube has as its vertices the 16 vectors (w, x, y, z), where each of w, x, y, and z can be either 0 or 1. This pattern continues up to the five-dimensional hypercube (with 32 vertices) and so on.

Fig. 3.7: I am lying a little in the main text. The image on the screen will be scaled if the screen is parallel to the table. If the screen is tilted, then the image will be altered by a projective transformation.

Fig. 3.9: 3D print by Bathsheba Grossman.

Fig. 3.11: The flashlight in these stereographic projection photographs has a "candle" feature. The lens can be removed, revealing a small LED that shines in all directions. This gives a wider spread that shows more of the shadow pattern. The smaller the light source, the sharper the shadows, so a small LED works well. Cellphone flashlights are also good.

To make it easier to get the positioning correct for these stereographic projection photographs, I taped the flashlight to a stick that is hanging down from a cross beam supported by a clamp stand on either side.

Fig. 3.12: The shadow pattern here is a simple grid, but almost any design is possible. All you need to do is cut the correct shapes out of the sphere to cast the shadows you want on the table. The basic procedure I follow is to start with the design made out of curves on the plane, and a spherical shell. Next, I make cones whose bases are the curves on the plane and whose vertices are all at the north pole of the sphere. Then I boolean subtract the cones from the spherical shell. This cuts out the correct windows from the spherical shell.

Thickening is a perennial problem in mathematical 3D print-ing. Usually we want to illustrate some one- or two-dimensional mathematical object, and these must be thickened somehow to be printed. Ideally, the grid pattern here would be painted onto the two-dimensional surface of a sphere, with zero thickness. Thickening becomes more of a problem here as we get nearer to the north pole of the sphere. If I had tried to go one further grid square up toward the north pole, then the rays of light coming from the north pole and hitting the upper and lower sides of the grid edge would have such a small angle between them that it would force the plastic to be too thin to 3D print.

Fig. 3.13: The formula for stereographic projection is remark-ably simple. For a sphere of radius 1 centered at the origin, the point (x, y, z) maps to the point $(x/(1 - z), y/(1 - z))$ on the plane $z = 0$. Assuming that our 3D printed spheres are radius 1, these shadow projections map to the plane $z = -1$. Moving the plane down from $z = 0$ to $z = -1$ doesn't change much. All it does is scale the projection up by a factor of two.

Fig. 3.14: 3D print and render by Saul Schleimer and me. There is a choice we made here in how to thicken up the edges of the hypercube. We could have projected the edge curves to three-dimensional space and then thickened them using tubes of some fixed radius around the curves. Instead, we thicken at the intermediate stage of the three-sphere, using tubes of fixed radius as measured in the three-sphere. The distortion of stereographic projection then makes it look like the tubes have varying thick-ness in three-dimensional space. This gives a more symmetric result, in the sense that the ratio of the width of a square face to the thickness of its edges is roughly the same for all of the faces of the hypercube. For further details on these models of four-di-mensional polytopes, see my article with Saul Schleimer, "Sculp-tures in S^3," which was published in *Proceedings of Bridges 2012: Mathematics, Music, Art, Architecture, Culture* (2012), 103–110.

Fig. 3.17: 3D print and render by Saul Schleimer and me.

Fig. 3.18: 3D print by Saul Schleimer and me.

Fig. 3.19: 3D print by Saul Schleimer and me.

Fig. 3.20: 3D print by Saul Schleimer and me.

Fig. 3.21: Render by Saul Schleimer and me. The width of the largest edge here is almost 30 times the width of the smallest edge. If we tried to 3D print this, we would have to scale the whole thing up so that the thinnest edge is above the minimum printable thickness. This would make the print very big and very expensive; instead, we only printed half of it.

Fig. 3.22: 3D print by Saul Schleimer and me.

Fig. 3.23: 3D print by Saul Schleimer and me.

Fig. 3.25: 3D print by Saul Schleimer and me.

Fig. 3.26: 3D print and photograph by George W. Hart (http://georgehart.com). Ignoring the colors, the symmetry type is 532.

Fig. 3.27: Render by Saul Schleimer and me.

Fig. 3.28: Render by Saul Schleimer and me.

Fig. 3.29: 3D print by Saul Schleimer and me. The "wrapping around" scheme here can be seen as a combinatorial version of the Hopf fibration.

Fig. 3.30: 3D prints by Saul Schleimer and me. Also see our article "Puzzling the 120-Cell" in *Notices of the American Mathematical Society* 62, no. 11 (2015), 1309–1316. A more detailed version of the article is available at http://arxiv.org/abs/1310.3549.

Fig. 3.31: 3D print by Will Segerman and me.

Fig. 3.32: Render by Will Segerman and me. Also see my article with Vi Hart, "The Quaternion Group as a Symmetry Group," which was published in *Proceedings of Bridges 2014: Mathematics, Music, Art, Architecture, Culture* (2014), 143–150.

Fig. 3.33: 3D print by Will Segerman and me. The four-dimensional symmetry group is the binary tetrahedral group.

Fig. 3.34: 3D print by Will Segerman and me. The four-dimensional symmetry group is the binary dodecahedral group. The astute reader may have noticed that while the six-limbed monkey (including the head and tail) works very well with the six faces of the cubical cells of the hypercube, the cells of the 24-cell have 8 faces and the cells of the 120-cell have 12 faces. We considered using animals other than the monkey with the appropriate numbers of limbs. However, 12-limbed animals are hard to come by, particularly animals with interestingly expressive faces. We decided to stick with the monkey and not make all possible connections between the cells. Even with fewer limbs, the 3D print is structurally robust. Fewer limbs also make it easier to see into the structure than it would be with all of the connections.

Fig. 4.2: 3D print by Geoffrey Irving and me. These hinged 3D prints come out of the printer as is. I did not link hundreds of tiny triangles together by hand. This means that there is a tricky problem in designing the print. We have to position the triangles in the printer in some layout, with the triangles already linked together. Triangles cannot be touching or overlapping with any others, because they would come out of the printer fused together. As I discuss in the main text, there doesn't seem to be an easy-to-describe (and so easy-to-code) layout for the triangles, and it is an open problem whether it is even possible to embed the {3,7}

Schläfli tiling in three-dimensional space out as far as we want or if there is some maximum possible radius.

This model is *combinatorial radius three*, meaning there is a vertex in the middle, surrounded by a ring of triangles, then another ring of triangles surrounds the first ring, and then a third ring surrounds the second. To position the triangles, we start off with a triangulated mesh for the {3,7} Schläfli symbol tiling in the Poincaré disk model. So the combinatorial structure is correct, and the triangles do not intersect each other, but the lengths of the edges are incorrect. We use this as the starting configuration for an iterative procedure that converges to a solution. We put virtual springs on all of the edges. The triangulated mesh buckles up and down away from the plane of the Poincaré disk as each spring attempts to change its length to a desired goal length, while detecting and avoiding collisions between the triangles. Once the edges have converged to their desired lengths, we copy hand-designed hinged pieces onto the triangles of the mesh, and we are done. Liu, Kim, Shiau and Séquin have also worked on numerical solutions to the layout problem. See their article "Large, Symmetric, '7-Around' Hyperbolic Disks" in *Proceedings of Bridges 2015: Mathematics, Music, Art, Architecture, Culture* (2015), 391–394.

Fig. 4.3: The circular paraboloid is given by the equation $z = x^2 + y^2$, the parabolic cylinder by $z = x^2$, and the hyperbolic paraboloid by $z = x^2 - y^2$. The geodesics and circles on the first two surfaces can be worked out knowing the length $L(t)$ of the segment of a parabola with x-coordinates between 0 and t. Or, rather, we want to use the inverse of this function, which can be approximated using Newton's method. The hyperbolic paraboloid is trickier and requires (numerically) solving a differential equation. (See Greg Kuperberg's answer to this MathOverflow question: http://mathoverflow.net/questions/25620/geodesics-on-a-hyperbolic-paraboloid.)

The curvature of a planar curve is a measure of how tightly it turns. For example, a circle of radius r has curvature $1/r$. So a circle with a small radius has large curvature and vice versa. Given a point p on a surface, consider the planes that go through p and contain the normal vector to the surface at p. As you rotate the plane around the normal vector, we see different curves of intersection between the plane and the surface, and these curves have different curvatures. The Gaussian curvature $K(p)$ of the surface at p is usually defined as the product of the maximal and minimal curvatures you see. These may have different signs, if the curves of intersection sometimes bend one way and sometimes the other. This corresponds to negative gaussian curvature. If the surface is smooth, then an alternative expression for K is given by the Bertrand-Diquet-Puiseux theorem, which states that, where $C(r)$ is the circumference of the geodesic circle of radius r around p on the surface. This then lets us think about gaussian curvature in terms of circumferences of circles, as we do in the main text.

Fig. 4.4: The circular paraboloid $z = x^2 + y^2$ has gaussian curvature $4/(1 + 4x^2 + 4y^2)^2$, while the hyperbolic paraboloid $z = x^2 - y^2$ has gaussian curvature $-4/(1 + 4x^2 + 4y^2)^2$. The only difference between the two formulas is the sign. By Pythagoras' theorem, $x^2 + y^2$ is the square of the distance from the origin, so the gaussian curvature for both of these surfaces depends only on this distance.

A torus with major radius (from the center of the hole to the center of the tube) R and minor radius (from the center of the tube to the surface) r can be parametrized by $(x, y, z) = ((R + r\cos\phi)\cos\theta, (R + r\cos\phi)\sin\theta, r\sin\phi)$. For this parametrization, the gaussian curvature at the point given by θ and ϕ is $(\cos\theta)/(r(R + r\cos\phi))$.

Fig. 4.5: Repeatedly subdividing triangles into four triangles subdivides each original edge into a number of smaller edges, in fact a power of two. The advantage of following this method is that it is always clear where to put each new vertex. Put a vertex at the midpoint of each edge and then move it out to be at the same distance from the center as the original vertices. But what if we want the number of smaller edges to be three or some other number that isn't a power of two? As far as I can tell, there isn't a standard way to choose where to put extra vertices in the middle of a triangle. For real-life architectural geodesic domes, the situation is further complicated. Rather than a mathematically perfect dome, approximations are used. These use a smaller number of different edge lengths, which makes it easier to manufacture the edges and organize building a dome.

Fig. 4.8: 3D print by Geoffrey Irving and me. We designed this in the same way as the print in fig. 4.2, except that here our virtual springs do not all have the same desired lengths. There are two different desired lengths: the two different lengths of the isosceles triangles. We chose these lengths so that they are related to each other in the same ratio as the corresponding geodesic lengths in the hyperbolic plane.

Fig. 4.9: The pseudosphere is a *surface of revolution*. That is, it is a surface that you can make by rotating a curve around an axis. (The torus in fig. 4.4 is another example.) The curve that we rotate around is the *tractrix*. Tie a string to a rock, put the rock at the point (1,0) on the plane, and stretch the string out so that your hand is at (0,0). Then start moving your hand up the *y*-axis. The path the rock takes as you drag it by pulling the string is the tractrix. It can be parametrized as $(x, y) = (1/\cosh t, t - \tanh t)$, for $t > 0$.

Fig. 4.10: For more on these triangle tilings, look up *Schwarz triangles*, or *Möbius triangles*.

Here's a puzzle for you to ponder. Stereographic projection of the sphere to the plane is very similar to the Poincaré disk model. Some of the "straight lines" (great circles) map to actual straight lines and others to circles, as you can see in this figure. But is there something like the Klein model for the sphere? Is there a projection of the sphere to the plane, which sends every great circle on the sphere to a straight line on the plane?

Fig. 4.12: Renders by Roice Nelson. There is an interesting comparison between the way Roice generated these images and the way Saul and I generated the 3D print in fig. 4.13. Saul and I used a *vector graphics* description of the shapes, which writes down coordinates of the endpoints of lines, the centers of circles, and so on. Roice used a *pixel graphics* description, deciding for each pixel of the image what its color should be. A 3D print is eventually described in terms of *voxels*: three-dimensional pixels. Either a voxel is filled with plastic or not. But the current standard formats for communicating with 3D printers consist of triangulated meshes, whose vertices are described in the same way as vector graphics, so it is more natural to use vector graphics. We also need to worry about the strength of a 3D printed object, so we could only go so far in cutting smaller and smaller windows out of the bowl. If we kept on cutting, at some point the connections between the rim of the bowl and the rest of it would become too weak.

Roice's pixel images have no such problems, and he can tile out as far as he wants. There is also a difference in the "direction" in which we generate the triangles. Saul and I start with a triangle in the middle and make copies, essentially by reflecting it across its geodesic sides. We keep reflecting outward to tile out to some distance. Roice also starts with one central triangle. Then, for each pixel, he has to decide what color it should be. If the pixel is inside of the central triangle it gets colored appropriately. If it isn't, then he repeatedly reflects the position of that pixel, moving it *inward* toward the central triangle. Once this position gets inside of a central triangle, the place where it lands together with whether it took an even or odd number of reflections to get there determines the color of the pixel.

Fig. 4.13: 3D print by Saul Schleimer and me. In reflecting triangles around, we also have to worry about not drawing more than one copy of a triangle in the same place. There are many different ways to reach some triangle from the center by doing a sequence of reflections, and it isn't so obvious how to make a list of sequences of reflections that both gets to every triangle (out to some radius) and doesn't repeat triangles. We used kbmag, a program by Derek Holt, which implements the Knuth-Bendix algorithm in order to solve this group theory problem.

Fig. 4.16: Render by Roice Nelson.

Fig. 4.21: 3D print by Roice Nelson and me. For more details on these 3D prints and on how to visualize all of the length three Schläfli symbols, see my article with Roice, "Visualizing Hyperbolic Honeycombs" (available at http://arxiv.org/abs/1511.02851).

Fig. 4.22: 3D print by Roice Nelson and me.

Fig. 4.23: 3D print by Roice Nelson and me.

Fig. 5.1: These flexible knots can be made out of string, although it is a little inelegant. You have to have a special part of the knot where the ends of the string are tied or otherwise connected together, while a mathematical knot doesn't have any special point on it. So using "pop-beads" is better. Pop beads are also produced in bulk using injection molding, at a much lower price than the 3D printed versions I designed here. As you may have guessed, however, I believe that just because one can 3D print something, one should absolutely 3D print it.

I cheated a little to make the crossings easier to see in these pictures, by shading the beads on the bottom of each crossing in postproduction.

Fig. 5.3: Yes, this is also the trefoil. Take the long strand on the left, lift it up and pull it over the rest of the knot toward the lower right. After laying it down again, you should get a distorted version of the trefoil in fig. 5.1.

Fig. 5.4: 3D prints by Keenan Crane and me. Keenan started with the coffee mug design, and then used an iterative method to deform the mug to the doughnut. This is analogous to simulating a soap film deforming from a high-energy to a low-energy state. For more details, see "Robust Fairing via Conformal Curvature Flow," by Keenan Crane, Ulrich Pinkall, and Peter Schröder, SIGGRAPH 2013 / *ACM Transactions on Graphics* 32, no. (2013).

Fig. 5.5: The vertices on the torus on the left are spaced so that the faces on the inside ring of the hole are all squares, as are the faces on the outside ring of the hole, making the shapes of the faces similar to the grid of squares on the right. Evenly spacing the vertices according to the usual parametrization of the torus in three-dimensional space $((x, y, z) = ((R + r \cos \phi)\cos \theta,$ $(R + r \cos \phi)\sin \theta, r \sin \phi))$ wouldn't have this effect, because it moves around a circle that goes through the hole at constant speed. In this model, the vertices are bunched up more in the middle and spaced out around the outside. The vertices are spaced according to a parametrization of the *Clifford torus* in the three-sphere, which I then stereographically projected to three-dimensional space. This parametrization of the Clifford torus is given by $(w, x, y, z) = (1/\sqrt{2})(\cos \theta, \sin \theta, \cos \phi, \sin \phi)$. In figs. 6.12 and 7.6, we saw two different ways to try to make a square flat torus in three-dimensional space. It turns out that the Clifford torus is a smooth, square flat torus, with zero gaussian curvature everywhere. So there is a simple solution to the problem of making a square flat torus in the three-sphere (and so also in four-dimensional space, which the three-sphere sits inside of).

Fig. 5.6: I added the red coloring on all three of these images in postproduction, after taking the photograph. At the time of writing, the current color printing technologies aren't particularly good for something long and somewhat flexible, like a knot. For the third image, I could have dyed the knot, although this wouldn't work for the second image because it needs more than only one color. I *could* have hand-painted the knot but that sounds too much like hard work to me.

Working digitally rather than physically has many advantages. I can undo mistakes, I can easily make copies, I can automate many of the boring bits, and so on. 3D printing lets me convert from digital to physical while retaining almost all of the advantages of the digital. But if I start painting things that all goes out the window.

Fig. 5.7: These torus knots are all parametrized on the Clifford torus by choosing a relative speed for θ and ϕ. So, for example, the (3,2) torus knot is parametrized by $(w, x, y, z) = (1/\sqrt{2})$ $(\cos 3t, \sin 3t, \cos 2t, \sin 2t)$. After stereographically projecting to three-dimensional space as in the figure, the (p, q) torus knot has symmetry type $22p$. It is also possible to stereographically project (from a different projection point) to get symmetry type $22q$.

Fig. 5.8: 3D prints by Ted Ashton, Jason Cantarella, Michael Piatek, and Eric Rawdon. For more details, see their article "Knot Tightening by Constrained Gradient Descent" in *Experimental Mathematics* 20, no. 1 (2011), 57–90.

Fig. 5.9: 3D prints by Ted Ashton, Jason Cantarella, Michael Piatek, and Eric Rawdon.

Fig. 5.11: The parametrization here came out of a collaboration with François Guéritaud and Saul Schleimer. As with the torus knots, this is a map into the three-sphere. Let $A(t) = \varepsilon \sin 4t$. Then the parametrization is given by

$$w(t) = (1 - A(t)^2)(\lambda \sin t - (1 - \lambda) \sin 3t)/(1 + A(t)^2)$$
$$x(t) = (1 - A(t)^2)(\lambda \cos t + (1 - \lambda) \cos 3t)/(1 + A(t)^2)$$
$$y(t) = (1 - A(t)^2)(2 \sin 2t)/(1 + A(t)^2)$$
$$z(t) = 2 A(t)/(1 + A(t)^2)$$

We chose the values $(0.25, 0.16)$ for the parameters (λ, ε).

The symmetry type of this knot in three-dimensional space is 2×. In the three-sphere (before stereographic projection), there are extra symmetries involving reflection across the equatorial two-sphere, for a total of eight symmetries. This parametrization has all possible symmetries of any parametrization of the figure-eight knot.

Fig. 5.12: The answer to the puzzle of whether there are borromean rings with four or more rings is, "yes." These are known as *brunnian links*. (Look them up online if you want to see some examples.) There are brunnian links with any number of rings.

Fig. 6.1: These are parametrized in the three-sphere and then stereographically projected to three-dimensional space. The parametrizations are variants of the parametrization of the *Round Möbius strip* from my article with Saul Schleimer, "Sculptures in S^3" in *Proceedings of Bridges 2012: Mathematics, Music, Art, Architecture, Culture* (2012), 103–110. In the article, we thicken up surfaces in the three-sphere (also see the commentary for fig. 3.14). Here, I thicken up surfaces in ordinary three-dimensional space. For the animation in fig. 6.6, there are parts of the surface that get relatively near to the projection point for stereographic projection, so there would be a large ratio between the thinnest and thickness parts of the surfaces. This is problematic. Although I could choose the thickness of the surface so that the thinnest part of the surface is above the minimum printable thickness, this would make the thickest part very thick and so expensive. In any case, the discussion in the main text is about the two-dimensional surfaces, so varying the thickness would be a distraction.

Instead of talking about two-sided versus one-sided surfaces, mathematicians usually talk about *orientable* versus *nonorientable* surfaces. For the examples in this book, all two-sided surfaces are orientable and all one-sided surfaces are nonorientable, but this is not always the case. A surface is orientable if it makes sense to talk about handedness—if a flatlander living on the surface could be consistently right-handed or left-handed. A Möbius strip isn't orientable, because a flatlander walking all the way around the strip would flip from being right-handed to left-handed or vice versa. Orientability is an *intrinsic* property of a surface, while the number of sides depends on what higher-dimensional space the surface sits inside of. It is possible to have a two-sided Möbius strip or a one-sided annulus, if you put them inside of a nonorientable three-manifold in the right way. In this book, we have mostly been thinking of surfaces as sitting inside of ordinary three-dimensional space, and here one-sidedness is the same as nonorientability. So I've stuck with the simpler-to-explain idea of the number of sides.

Fig. 6.2: I made these with the by-hand method, using the design tools in my design software, Rhinoceros. It is surprisingly difficult to come up with reasonably canonical shapes in three-dimensional space for surfaces without boundary and genus bigger than one.

Fig. 6.3: 3D print and photographs by Bathsheba Grossman. Bathsheba designed this entirely in Rhinoceros, using the Sporph command to flow the thickened hexagonal grid along a two-dimensional guide surface.

Fig. 6.4: 3D print by Saul Schleimer and me. Saul and I call this the *Round Möbius strip*. As we see in the main text, this is a Möbius strip (with an extra hole) in disguise. The boundary of an ordinary Möbius strip, as seen in fig. 6.1, is an unknotted loop in space. So, topologically it is the same as a circle, but geometrically it isn't a circle; it isn't "round." For the *Round Möbius strip*, however, this boundary is geometrically a circle.

Fig. 6.6: The only thing changing in this animation is the position of the projection point for stereographic projection: it travels around one quarter of a great circle of the three-sphere as we move from the left to the right in the figure.

Fig. 6.8: 3D print and photograph by Oliver Labs (http://www.MO-Labs.com). Oliver has produced many beautiful prints of surfaces that are defined algebraically, as the set of solutions to some equation or equations. Felix Klein (of the bottle, the model of the hyperbolic plane, and the quartic) lived in the late 1800s and early 1900s. He promoted the use and supervised the construction of numerous mathematical models, many of them of algebraic surfaces, made from plaster and other materials. (Unfortunately, they didn't have 3D printers at the time.) He argued for the use of models to aid in intuition and argued against too much abstraction of mathematics away from tangible reality.

The Clebsch diagonal surface was first described by Alfred Clebsch in 1871. This object lives in three-dimensional complex projective space and is described by the equation $w^3 + x^3 + y^3 + z^3 = (w + x + y + z)^3$. This is a very symmetrical thing in its native setting, but as usual, we lose most of the symmetry when we project it down to real three-dimensional space. A word of warning: In three-dimensional complex projective space, this is a *complex surface*. If you zoom in on a small part of the Clebsch diagonal surface in its native setting, it looks like four-dimensional space. (It has two *complex dimensions*, each complex dimension counts for two real dimensions.) The projection to three-dimensional space cuts the number of dimensions in half, so that the complex surface becomes a *real surface*, what I've been calling a *surface* in this book.

The lines drawn on the surface are important: it turns out that any smooth cubic algebraic surface has exactly 27 lines on it, although for many surfaces it isn't possible to see all of them together in a single projection to three-dimensional space. For the Clebsch diagonal surface, this is possible, and they are drawn on the print as raised lines.

When we project to three-dimensional space, the surface we get extends outward infinitely (the lines extend forever). This would be somewhat expensive to print in its entirety, so Oliver had to cut it off in some way. He chose a cylinder to do this—other choices (e.g., cutting off with a sphere) would give a different look to the print. The original plaster models from a century ago were cut off with both a cylinder and a horizontal plane. They were also designed as a solid mass of plaster, with the surface illustrated as . . . well, the boundary surface of the mass of plaster. In contrast, with 3D printing, we can have a thin membrane illustrating the surface, so that we can see both sides of it.

This surface has Euler characteristic −5, 2 sides, and 1 boundary loop, so it is the same as the genus three surface with a hole punched out of it.

Fig. 6.9: Sculpture and photograph by Carlo H. Séquin, University of California, Berkeley. The geometry comes from Carlo's *Sculpture Generator 1*, details in his article "Virtual Prototyping of Scherk-Collins Saddle Rings" in *Leonardo* 30, no. 2 (1997), 89–96. The generator program is also available at Carlo's website (currently at http://www.cs.berkeley.edu/~sequin/GEN/Sculpture Generator/).

This surface has Euler characteristic −6, 1 side, and 3 boundary loops, so it is the same as a sphere with nine holes punched out of it, and Möbius strips glued onto six of the resulting boundary loops.

Fig. 6.10: 3D print by Saul Schleimer and me. Just as we parametrized the torus links in the three-sphere, the Seifert surface for the link also lives most naturally in the three-sphere. Then, as usual, we use stereographic projection to get it into three-dimensional space. The (p, q) torus knot, or link, can be described by the equation $w^p + z^q = 0$ in the three-sphere, seen as sitting in two-dimensional complex space. The Seifert surface can then be described as a *Milnor fiber*, by the equation $\arg(w^p + z^q) = 0$. The shape is parametrized, using certain fractional automorphic forms, by a map from the hyperbolic plane to the three-sphere.

This surface has Euler characteristic –3, 2 sides, and 3 boundary loops, so (astonishingly) it is the same as a torus with three holes punched out of it. I don't know of a good way of seeing this directly though.

Fig. 6.11: 3D print by Bathsheba Grossman. Bathsheba's sculpture is based on output from a program called SeifertView, developed by Jarke J. van Wijk and Arjeh Cohen. See their article "Visualization of Seifert Surfaces" in *IEEE Transactions on Visualization and Computer Graphics* 12, no. 4 (2006), 485–496. Bathsheba also used a program by Ken Brakke called Surface Evolver to further alter the shape of the surface. She then thickened the surface and added the pattern of holes. Her algorithm relates the density of the holes to the curvature of the surface.

This surface has Euler characteristic –3, 2 sides, and 3 boundary loops. Intrinsically, it is the same surface as the (3,3) torus link Seifert surface: a torus with three holes punched out of it.

Fig. 6.12: I haven't been able to definitively track down who first came up with this design. Joseph O'Rourke directed me to an article by V. A. Zalgaller, "Some Bendings of a Long Cylinder" in *Journal of Mathematical Sciences*, 100, no. 3 (2000): 2228–2238 (translated from a 1997 article in the Russian journal *Zapiski Nauchnykh Seminarov POMI*). This article has some similar constructions, so we think that Zalgaller may also have come up with the design shown in this figure. I first learned of it at http://www.mathcurve.com/polyedres/toreplat/toreplat.shtml.

Is there a simple way to make a square, hinged flat torus, like the print in this figure but that folds out into a square? There is apparently a very complicated way to do this, requiring an enormous number of hinged pieces, described in an article (in Russian) by Yu. D. Burago and V. A. Zalgaller, "Isometric

Piecewise-Linear Immersions of Two-Dimensional Manifolds with Polyhedral Metrics into $\mathbb{R}^3$" in *St. Petersburg Mathematical Journal* 7, no. 3 (1996), 369–385.

Fig. 6.13: 3D print by Saul Schleimer and me. The Klein quartic most naturally lives in two-dimensional complex projective space, as the set of solutions to the equation $x^3y + y^3z + z^3x = 0$. We used a parametrization of the surface from the hyperbolic plane due to Srinivasa Ramanujan and described by Gilles Lachaud in his article "Ramanujan Modular Forms and the Klein Quartic," which was published in *Moscow Mathematical Journal* 5, no. 4 (2005), 829–856. We mapped the surface down to three-dimensional space using (3,3) bihomogeneous polynomials for the three coordinates, chosen in a way that would preserve the tetrahedral 332 symmetry, and taking real parts. This still gives a very large space of possible ways to map the surface into three-dimensional space. We chose a map that balances two things: First, we didn't want the surface to get close to crashing into itself. Second, we wanted to minimize distortion. The "buttons" are circular in the hyperbolic plane but get squashed into approximate ellipses in three-dimensional space. We tried to reduce the eccentricity of these ellipses. To find this map, we used a hill-climbing algorithm to navigate the space of possibilities, optimizing a function that combines a measure of how close the surface is to crashing into itself with a measure of the distortion. This optimization introduced the "twisting ripples" in the surface.

Fig. 6.14: This makes it easier to see how to deform the surface in fig. 6.13 to make it clear that it is the same as a sphere with three handles added. A good intermediate step between the two is the surface you get by thickening up the edges of a tetrahedron. We can see the genus of this "thickened-tetrahedron-edges surface" by building it up from the different edges. Three thickened edges connected together to make a triangle is a torus with genus one. Adding two more of the tetrahedron's edges adds a handle to the torus and so gets you to genus two. And adding the last thickened edge adds one more handle and gets you to genus three. We can get the thickened-tetrahedron-edges surface from the surface in fig. 6.13 just by smoothing it out a bit, removing the ripples. Each pair of pants corresponds to a (thickened) vertex of the tetrahedron together with three (thickened) half edges.

We could have drawn the (7,3,2) triangle tiling pattern on just a thickened-tetrahedron-edges surface, rather than on the much more rippled surface in fig. 6.13. However, we think that the distortion of the map would have to be much worse. The buttons would have been much more squished and the triangles much skinnier—to the extent that it would have been difficult to print in this style.

Fig. 6.15: Render by Roice Nelson. "Uniform geometry" is not a standard term, and I am being deliberately vague with definitions. For closed surfaces, I mean "constant curvature," although this doesn't generalize particularly well to higher-dimensional manifolds. A geometry in which you can't "fall off the edge" is called a *complete* geometry, and the precise more general definition of uniform geometry I'm thinking of is described by something called a *(G,X)-manifold* (see William Thurston's book, *Three-Dimensional Geometry and Topology*, p. 125).

Fig. 6.16: I changed the pattern of holes in the torus here (in comparison to fig. 5.11) so that the inside knot is visible through a single hole.

Fig. 6.17: As with many of the (apparently) colored prints in this book, I added the colors here in postproduction—after taking the photograph. At the time of this writing, color 3D printing technology isn't yet good enough to do the design justice.

Fig. 6.18: This uses the same parametrization as in fig. 5.11 again, but with a different geometry built along it. I added by-hand the parts of the surface that bridge across the "clasp" of the knot, following the two edges through the manifold. This way of cutting up the complement of the figure-eight knot and putting hyperbolic geometry on the pieces is described in Thurston's notes on *The Geometry and Topology of Three-Manifolds* (available from http://library.msri.org/books/gt3m/), chapters 1 and 3. For more on hyperbolic knot complements, tabulating knots, and why so many knots have hyperbolic complements, see "The First 1,701,936 Knots," by Jim Hoste, Morwen Thistlethwaite, and Jeff Weeks, in *The Mathematical Intelligencer* 20, no. 4 (1998), 33–48.

Fig. 6.19: The curve I used here is the intersection of a hyperbolic paraboloid with a sphere symmetrically centered on it. Beware that the hyperbolic paraboloid is *not* a minimal surface, so the soap film shape is only a close approximation of the hyperbolic paraboloid.

Fig. 6.22: 3D print and photographs by Bathsheba Grossman. Bathsheba designed this in a similar way to *Borromean Rings Seifert Surface* (as in fig. 6.11) but starting with an initial surface produced by Mathematica.

Fig. 6.23: This model is given by the implicit equation $\sin x \cos y + \sin y \cos z + \sin z \cos x = 0$. The shape is a close approximation to the true gyroid, but is not itself minimal.

Fig. 6.25: A smaller frame around a smaller part of the gyroid

surface would, however, find the right shape. It seems that too much genus in a surface makes it unstable as a soap film in the real world.

Fig. 7.1: 3D print: Hyphae lamps by Nervous System (http://nervo.us.). Photograph by Jessica Rosencrantz.

Fig. 7.2: 3D print by Marco Mahler and me. 3D prints structured like trees (without loops in them) often have weak points where they can break, because there are no cross supports. A mobile gets around this problem by flexing freely. It isn't easy to apply a lot of force to any particular point.

Fig. 7.4: 3D print by me, based on an idea by Geoffrey Irving. For more details on this construction, see our article "Developing Fractal Curves" in *Journal of Mathematics and the Arts* 7 (2013), nos. 3–4, 103–121.

Fig. 7.5: Sculpture and photograph by Carlo H. Séquin, University of California, Berkeley. Carlo describes the geometry in his article "Analogies from 2D to 3D—Exercises in Disciplined Creativity" in *Bridges: Mathematical Connections in Art, Music, and Science* (1999), 161–172.

Fig. 7.6: This is a square flat torus embedded in euclidean space. As I mention in the main text, this is slightly smoother than a polygonal torus. It is in the C^1 class of surfaces; roughly speaking it has enough smoothness to differentiate once but not twice. Nash and Kuiper proved the existence of this shape in the 1950s, but nobody knew what it looked like until the work of Borrelli, Jabrane, Lazarus, and Thibert. For more details, see their article "Flat Tori in Three-Dimensional Space and Convex Integration" in *Proceedings of the National Academy of Science* 109, no. 19 (2012), 7218–7223. This model was printed by GI-Nova (http://www.aip-primeca-ds.net/-GI-Nova-.html).

Fig. 7.7: 3D print by Craig S. Kaplan and me. See Craig's article "Semiregular Patterns on Surfaces" in *NPAR 2009: Proceedings of the 7th International Symposium on Non-photorealistic Animation and Rendering*. His work uses the surface parametrization techniques developed in "Spectral Surface Quadrangulation" in SIGGRAPH 2006, by Dong, Bremer, Garland, Pascucci, and Hart. The shape here is the Stanford bunny, a standard 3D graphics test model developed by Greg Turk and Marc Levoy.

Fig. 7.8: 3D print and photo by M. Oskar van Deventer. After I finished writing this book but before publication, a physical 22 × 22 × 22 Rubik's cube was constructed (see http://www.thingiverse.com/thing:1267855).

Fig. 7.9: 3D print and photo by M. Oskar van Deventer. The first gearing design along these lines seems to be due to F. E.

Lindsay. See his article "An Interesting Gear" in *The Engineer* 135 (1923), 660. Oskar added the idea of using *herringbone gears*, in which two helical gears with opposite twist directions are combined into a single gear. Here, the herringbone-style gears can demonstrate both ordinary gearing and the paradoxical "same direction" gearing.

Fig. 7.10: 3D print by Saul Schleimer and me. For further details, see our article "Triple Gear" in *Proceedings of Bridges 2013: Mathematics, Music, Art, Architecture, Culture* (2013), 353–360.

Appendix B: How I Made These Models

Technology moves quickly and whatever I write here may be obsolete advice in a few years. But I can say something about how I make 3D models at the moment.

My main 3D design program is Rhinoceros, developed by Robert McNeel & Associates. There are many different 3D design programs out there, for many different uses and based around different ways to represent 3D objects. Rhinoceros seems to be the best for mathematical designs. Bathsheba Grossman recommended that I try Rhinoceros when I was first getting into 3D design. It makes it very easy to do "ruler and compass" constructions, intersecting lines and other curves with one another and with surfaces and so on. It has many alternate ways to generate an object. You can of course generate a sphere by giving its center and radius, but you can also tell Rhinoceros to make a sphere that goes through four chosen points. One further piece of evidence in favor of Rhinoceros for mathematical uses: Edmund Harriss (mathematician and mathematical artist, he often works with laser cutters and CNC routers) tells me he performed the following experiment several years ago. He downloaded the trial versions of many of the CAD programs and challenged himself to make a stellated dodecahedron from scratch, without looking at any tutorials or other help. Using Rhinoceros, it took him 30 minutes. Every other program took more than two hours.

At present, 3D printers generally take as input a *mesh*, which is a surface made out of flat, euclidean triangles and/or quadrilaterals. They then use the mesh to determine whether each point in space is inside of the mesh, and so should be made out of plastic (or whatever material is being used), or is outside of the mesh, and so should be made out of air. For this to work, the mesh should be a two-sided surface without any boundary loops, and which does not crash through itself.

A mesh can be stored in a computer file in a few different ways. Two of the most commonly used are *stl* files and *obj* files. A standard stl file is essentially a list of triangles, with each triangle given by storing numbers for the x, y, and z coordinates of its corners. An obj file is a bit more general and can describe quadrilaterals and triangles as well as color and texture information.

Many 3D design programs work directly with meshes; however, Rhinoceros works primarily with *NURBS surfaces*. The acronym "NURBS" stands for non-uniform rational B-spline, a not-so enlightening name. They describe surfaces, either organically curved or with simpler geometric shapes, using polynomials. NURBS surfaces are generalizations of *Bézier curves*, which are used in two-dimensional vector graphics programs, such as Adobe Illustrator, to represent curves. Just as with Bézier curves in Illustrator, the user interface of Rhinoceros doesn't require you to work with the raw polynomials, instead there are lots of higher-level intuitive tools for working with them. Once a NURBS surface design is finalized, Rhinoceros can convert it into a mesh for sending it to a 3D printer.

I often use Rhinoceros alone (the by-hand method) if the shape is relatively simple. For example, I made all of the comma symmetry spheres from chapter 1 using only the user interface. Patches of NURBS surfaces can be joined together to make closed surfaces, and then further shapes can be obtained using boolean operations. I made the comma spheres, for example, by boolean subtracting extruded comma shapes from spherical shells.

For more complicated shapes, Rhinoceros also has a Python scripting interface, which lets me write code that generates an object. I made many of the designs in this book entirely with Python scripts, including the polyhedra in chapter 2 and the polychora in chapter 3. For the polyhedra, it made most sense to produce a mesh directly because the models are naturally made from flat polygons. For the polychora and many other curvy objects, I wrote code to generate NURBS surface patches that fit together to make the object. My usual process is to first figure out a parametrization of each patch of surface I need. Using the parametrization, I then plot a grid of points on the patch and fit a NURBS surface patch to those points.

Often a script uses input from some outside source, for example lists of vertices of polyhedra or polychora. Some of these sets of data came from *Wikipedia*, and others came from databases within the mathematical software package Mathematica. Mathematica is often also helpful for making symbolic calculations, the results of which can then be turned into Python code. For many of my collaborations with Saul Schleimer, the parametrizations themselves were very complicated, requiring mathematical functions not available within Rhinoceros's implementation of

Python. For these, we used the Python library mpmath. We also used the mathematical software system Sage for symbolic and algebraic computations.

For my usual process to work, however, I need a parametrization of surfaces used to define an object. Sometimes I don't have a parametrization to use, either because a shape is defined only implicitly or it can only be generated by an iterative method.

For implicitly defined surfaces, Mathematica is often a good option. It can export a mesh, which I can then bring into Rhinoceros for further work. Mathematica is useful if you are thinking about getting started in mathematical 3D printing, particularly if you already have it through work or school. Mathematica's RegionPlot3D function produces output that can be exported as an stl file and fed directly to a 3D printer.

I don't have good tools myself for generating surfaces by iterative methods. This is what having wonderful collaborators is for.

Some of the designs are hybrids, for which I used both Rhinoceros's user interface and scripts. Sometimes a model generated by scripts needs a little cleaning up afterward by hand. The hinged surfaces in figs. 4.2 and 4.8 are hybrids in a different way. As I mentioned in appendix A, I made the hinged triangles by hand and used a script to move them to the correct positions. The correct positions were calculated with code written by Geoffrey Irving.

I have been using yet another kind of hybrid idea more and more recently. Many of my designs are relatively simple grids, distorted using a relatively complicated parametric function (for example, the twist surfaces in fig. 6.1). For a long time, my process was to generate the input grid geometry using code and then feed it through the parametric function to make the final object. It took me an embarrassingly long time to realize that I could make the input geometry by hand in Rhinoceros, then have a script read it in and push it through the parametric function. This makes it much easier to tweak a model. For example, the raised lines along the middle of the twist surfaces would have been annoying to make by writing code but were very easy to add to the input geometry by hand. The monkey sculptures in figs. 3.31, 3.33, and 3.34 grew out of this strategy: My brother, Will Segerman, designed the monkey in a cube, as shown in fig. 3.32. This is the input geometry, which gets sent through eight different parametric functions to make the eight monkeys in fig. 3.31. We used similar inputs for the other two monkey sculptures.

After I have generated the geometry of a design and converted it to a mesh, there may still be some final steps to do. Sometimes the number of polygons in the mesh needs to be reduced, or the mesh needs to be checked to make sure that its walls are thick enough to be printed. Overlapping meshes may need to be boolean unioned into a single connected mesh, or the mesh may otherwise need to be repaired in some way. I use various tools for these kinds of jobs, including MeshLab, OpenFlipper, and netfabb Basic (also see https://netfabb.azurewebsites.net). For offsetting a mesh to hollow it out, I use a command line utility written by Tobias Sargeant.

I used the online 3D printing service Shapeways for all of my 3D prints in this book. Almost all of these were made in "White Strong & Flexible" material, with the exceptions of figs. 3.24, 3.25, and 4.4 (made with "Full Color Sandstone" material) and fig. 5.4 (made with their "Porcelain" material).

Index